LORENZO LORRAINE LANGSTROTH
AT EIGHTY YEARS OF AGE

TO THE MEMORY OF
LANGSTROTH

Published in the United States of America
First published 1942 by Cornell University Press
under the title The Life of Langstroth

Reissued 1976 with Foreword by Roger A. Morse

Republished under licence by Northern Bee Books, 2023
Scout Bottom Farm, Mytholmroyd, Hebden Bridge, HX7 5JS (UK).

ISBN: 978-1-914934-72-8

Almost all the hives in Greece today are 10 - frame Langstroths.
They differ from other versions of Langstroth hives by having
permanently fixed floors. *(John Phipps, Greece)*

AMERICA'S MASTER OF BEE CULTURE

The Life of L. L. Langstroth

BY FLORENCE NAILE

EDITED, WITH A FOREWORD AND AN INTRODUCTION, BY
EVERETT FRANKLIN PHILLIPS
AND A NEW FOREWORD BY
ROGER A. MORSE

Northern Bee Books

LORENZO LORRAINE LANGSTROTH

FOREWORD, 1976

FEW industries can point to one man, or pinpoint a single date in their histories, and say, "This was the beginning." The beekeeping industry is an exception, for the Reverend Lorenzo Lorraine Langstroth was undoubtedly the Father of American Beekeeping; and in his journal we can read how Langstroth discovered the principle of bee space and made modern beekeeping possible. The journal, now a part of the Everett Franklin Phillips Memorial Beekeeping Library at Cornell University, was an important source for the present book.

This volume is a new printing of Florence Naile's biography, which was published by Cornell University Press in 1942 under the title *The Life of Langstroth*. Since the original book is itself a historical document in the history of beekeeping, it has been reprinted exactly as it first appeared.

The mid-1800's were exciting years for American agriculture. New lands were being opened in the West, and a great revolution in farm machinery and equipment was taking place. Langstroth and his invention of the movable-frame hive were part of this movement, a movement which freed men from the soil and allowed them to follow other activities. These changes came at a time when over half the people in our country lived on farms and were busy producing food both for themselves and for a few others; today one American farmer grows the food for more than fifty people.

At the time of Langstroth's discovery and invention the annual per capita consumption of sugar in the United States was approximately 50 pounds; today it is 140 pounds. In 1851 consumption was limited by production, and production in turn was limited by a lack of machinery and mechanization. No doubt sweets were as much in demand then as they are today. Since the results of Langstroth's studies enabled apiculturists to increase honey production by at least 500 per cent— and honey then commanded a price equal to that of butter—one can quickly grasp the financial rewards gained by the early beekeepers. Tributes to Langstroth by successful beekeepers were common throughout his lifetime.

It is possible that Langstroth's greatest contribution to beekeeping was his book, *Langstroth on the Hive and the Honey-bee, a Bee Keeper's Manual*, first published in 1853. Langstroth not only revealed his discoveries to the world, but also, and more importantly, gave sound, practical advice on bee management. Without his writings the world might have known little about the principle of bee space and its implications for beekeeping. Late in his life, and after his book had gone through several editions, it was revised by Charles Dadant. Dadant and Sons, sellers of beekeeping equipment, published the twentieth edition in 1975.

In the present volume, Miss Naile's chapter "The Golden Bee Imported" is of special interest. Since certain bees were known to be better producers, Langstroth was concerned that American beekeepers have the best bees. Bees from northern Italy, often called "golden Italians," were brought to this country by Langstroth and others. At that time the rearing of queen bees was a little-known art. Langstroth's success in rearing daughters of these queens demonstrated his ability as a beekeeper. Even today, queen rearing is considered one of the more difficult of beekeeping practices.

It appears that never has there been so much interest on the part of so many in keeping bees as now. The honey bee is a producer of food and the single most important pollinator of our fruit, nut, and seed crops; the bee also contributes to a better understanding of science and biology, mainly because bees can be kept in ordinary and glass-walled observation hives, where their behavior may be studied. Langstroth's inventiveness and industriousness, so well portrayed by Florence Naile, has made it possible for interested persons to pursue beekeeping as a vocation, as an avocation, and in the interest of science. This volume tells how it all came about.

ROGER A. MORSE

ITHACA, NEW YORK
April, 1976

FOREWORD, 1942

SINCE the name of Langstroth has been honored for many years in all parts of the world where bees are kept, it may seem strange that somebody did not long ago write a book about his life and work. That omission can hardly have been due to indifference among his disciples. More probably some of them, contemplating such a task but knowing how laborious it would be, have hesitated to undertake it for fear of not doing the subject justice.

Thanks are therefore due Miss Naile for what she has accomplished, proving as it does that Langstroth did indeed deserve a biography. Hers has been a labor of love. It will be evident to her readers that she has given generously of time and effort to searching out the facts of this man's life and of his contributions to the science and practice of bee culture.

There is another reason for thanking Miss Naile. She has promised that whatever royalties might come to her as author shall instead go into a memorial endowment fund, already established, and be used for promoting studies in the field which Langstroth made his own. This Langstroth Memorial Fund, held in trust and invested by the board of trustees of Cornell University, is the means of maintaining and building up at this university a library of beekeeping. This library is itself in part an interesting memorial of Langstroth. Here is his personal Journal, the day-by-day record of his observations and discoveries in the apiary. Here is his letter-press book, with copies of many letters relating to his work. By good fortune this library has obtained a number of books which were once his own. It hopes to acquire more of them. A list of many that he is known to have owned will be found in the Appendix.

It is a privilege to sponsor this book. In expressing appreciation of it I venture to speak for beekeepers everywhere.

E. F. PHILLIPS

ITHACA, NEW YORK
February, 1942

PREFACE

THE story of a life of more than eighty years of love and usefulness and wisdom cannot be told in one small volume; and this work is attempted only to fill in some measure the great need that was expressed in the *American Bee Journal* on October 3, 1895: "The beekeeping literature of this country will remain incomplete as long as we are without a fair-sized volume giving the life and times of L. L. Langstroth. The biography of that noble man would be read with avidity by the beekeeping fraternity. It is with pleasure that beekeepers receive the announcement that he will again resume his literary labors. Would it not be well to secure from him while he is yet with us such data as would be necessary to complete his biography?" Three days later came the news of Mr. Langstroth's death, and the feeling of beekeepers was well expressed in these words of Professor A. J. Cook, of Michigan State College: "With thousands of others all over our great country, I bowed my head and heart with grief and sorrow that I should see the kindly face and hear the sympathetic voice no more."

Every movable-frame hive in the world tells mutely the story of his genius, yet beekeepers of today know almost nothing of the man himself, who toiled that they might reap.

In obtaining material for this biography, the author gratefully acknowledges indebtedness to the following persons: The grandchildren of Mr. Langstroth, sons and daughters of Mrs. Anna Langstroth Cowan, for the privilege of using his Journal, now donated to the Langstroth Memorial in the Cornell beekeeping library, for many of his personal letters, and for much other valuable information: Dr. Lovell Langstroth, San Francisco, California; Mrs. Bertha Langstroth Gant, Matamoras, Mexico; Miss Agnes Dunn, Philadelphia; the Reverend E. Victor Bigelow, pastor of South Church, Andover, Massachusetts; the Reverend A. P. Pratt, pastor of the Second Congregational Church, Greenfield, Massachusetts; Miss Jane Carpenter, Abbot Academy, Andover, Massachusetts; Mr. Marion T. Phillips, Alumni Registrar, Yale University; the Reverend William Mc-Surely, Miss Elizabeth McFarland, Mr. and Mrs. Llewellyn Bonham, Mrs. W. F. Brown, Miss Florence Brown, Madam Emily Hughes, and Miss Elizabeth McCord, Oxford, Ohio; Miss Margaret Linn, Hillsboro, Ohio; Mrs. Elizabeth Fisher, Frankfort, Ohio; Mr. Ph. J. Baldensperger, Nice, France; Mr. C. P. Dadant, former editor of the *American Bee Journal,* Hamilton, Illinois; Mr. E. R. Root, President of the A. I. Root Company, Medina, Ohio; Parker Dodge, Esq., Washington, D. C.; Miss Edith Auch and Dr. W. E. Dunham, Columbus, Ohio; the late George W. York, Seattle, Washington; also to Baldwin's *Annals of Yale College,* the Historical Society of Pennsylvania, the Maryland Historical So-

ciety, the Historical Department of Iowa, the Ohio
Archaeological and Historical Society, and to old files of
Gleanings in Bee Culture and the *American Bee Jour-nal.*

Especially to Dr. E. F. Phillips, Professor of Apicul-ture in Cornell University, the author would express
sincere thanks for his kindness in editing this volume,
and for valuable help and suggestions in the preparation
of the manuscript.

FLORENCE NAILE

COLUMBUS, OHIO
April, 1941

CONTENTS

ILLUSTRATIONS

INTRODUCTION

MODERN beekeeping had its inception in L. L. Langstroth's invention, and the story of his life shows that for many years the growing industry was wisely guided by his counsels. To suggest something of the superstructure built on this foundation, it seems admissible to present a brief survey of the accomplishments in beekeeping, and to note some of the trends of intervening years. It would be possible to present statistics of greater or less validity by states and countries on the status of beekeeping, but perhaps a better plan is to recount a few of the stages through which the industry has passed, to mention some of the later problems, to acknowledge errors, to suggest wherein seems to lie a brighter future. The record is not always worthy of the impetus which Langstroth gave this branch of agriculture.

To place monetary value on the industry which began with the work of Langstroth is almost sacrilege, since his first purpose was humanitarian, yet his life work was devoted to the practical establishment of honey production, and one cannot determine the degree of success attained without recognition of the wealth created by the care of bees. This is creation rather than exploitation, for without the bees this wealth would be lost to man. Throughout all the world, there are tens of thousands of persons who produce honey for human use. In the United States and Canada, where his influence was most direct, these persons perhaps number almost a million, which suggests how widely the Langstroth invention has brought financial benefit and mental and spiritual uplift. The honey produced by this army of beekeepers is shared by a far vaster number, and the consumers of honey are likewise beneficiaries of the work of a great and good man.

Pollination of fruits and forage crops depends on insect visits, and the myriads of colonies of bees provide insurance of the adequacy of this benefit to man through the service to plants. Here the creation of wealth far exceeds that from the honey crop, and becomes so vast that one dares not even estimate its amount. People have become so accustomed to this symbiosis, or mutual support of life, that the benefit is often overlooked, even by those whose very livelihood depends upon it. A striking demonstration of the service of honeybees was presented by the results of their neglect during the first World War in Central Europe, where the conduct of war became more important than the peaceful pursuit of beekeeping. Multitudes of bees were neglected and left to perish during the absence of their owners at the front, and such losses were incurred from inadequate pollination that, at the close of the war, strenuous even though misguided steps were forced on governments to salvage beekeeping as a basis for pollination. Those who most directly benefit from the presence of bees often continue practices which are detrimental to their insect friends.

During the years interesting changes in beekeeping practice have occurred. Not all locations have proven equally suitable for the development of honey production on a commercial scale. Where soil and climate combine to provide conditions most favorable to the abundant secretion of nectar from important floral sources, beekeeping has been developed on a scale adequate to provide livelihood for those who engage in no other work, and in these favored areas the industry has come to be so firmly established that it has weathered a decade of economic distress more soundly than other more conspicuous human endeavors. In those locations where nectar secretion is less abundant, or where the abundance of nectar is not so well timed with regard to the greatest colony strength, there are fewer efforts to engage in beekeeping as a specialty, but in these areas the industry is nevertheless important. Far more people keep bees in the less commercialized areas, and though the scale of operation is smaller the aggregate honey crop is usually not less than in an equal area which is more suitable for larger individual apiaries. In all these areas alike, the bees continue their good service in pollination. This distinction between commercial and non-commercial beekeeping areas is a development of fairly recent years and is one which Langstroth could scarcely have foreseen, since he did not keep large numbers of colonies, and since at no time except during his residence in Oxford, Ohio, did he keep bees in a location now considered quite favorable for beekeeping on a large scale. Even Oxford is not the best such location in Ohio.

Because of the facility with which bees can be handled with movable frames, much progress has been possible in a study of their behavior. Facts have been learned which to Langstroth were doubtless unsuspected. Far more is now known about the cause and control of swarming, and about the activities of bees during the broodless period of winter, and at last there is some basis of knowledge for maintaining conditions in the hive which will permit the colony to produce a maximum crop of honey.

The question of the right size and shape of frame and hive was once a burning problem among beekeepers. In earlier meetings of beekeepers, hours were spent in comparing hives, it being then generally assumed that the hive is the most important factor in the production of a good honey crop. In earlier days most beekeepers made their own hives and made them as they wished. With the development of factory methods it became too costly to make hives of different types, and the number of varieties was reduced. While some manufacturers have advocated the use of particular types, in general they have made such hives as were preferred by purchasers. After costly trials and eliminations, North American beekeepers have come to use two sizes of frame. A great majority of them use frames of almost the exact dimensions advanced by Langstroth in his first edition, with an inconsequential change in length of frame arising from an inadvertent mistake on the part of the manufacturer. There are some beekeepers who still prefer deeper frames, first used by Moses Quinby and later ably advocated by the Dadants. While frames of the Quinby-Dadant dimensions are extensively used in several European countries, the Langstroth dimensions are preferred by most Americans. It is an implied tribute to Langstroth that the frame dimensions which he devised have come to be so widely used. Interestingly enough, however, the basis for the present use differs from that of his day. Beekeepers have come to recognize that the skill of the operator is more important than the dimensions of hive or frame. Formerly a beekeeper always placed the blame for a poor crop on the hive, or the nectar sources, or the weather, but today there are a few beekeepers who are wise enough to admit that sometimes the fault lies with the operator. The Langstroth hive was devised for the production of comb honey in bulk and has since been used for the production of both section comb honey and extracted honey. Now that it has been discovered that a strong colony needs more room in the hive than was formerly provided, with far more space for stores during periods of adversity, advanced beekeepers use a hive composed of two stories of the Langstroth dimensions, the largest hive in use, and the best hive combination yet devised.

Beekeepers have thus taken frame dimensions devised for quite different conditions and have adapted them to use in a more advanced honey production.

Langstroth's name, strangely enough, is not commonly applied to the frame which he devised. Julius Hoffman constructed a device for keeping frames well spaced in the hive, especially useful in the production of comb honey. Because of this minor modification his name is usually applied to the frame in common use. However serviceable Hoffman's device may have proved, the American beekeeper is not justified in applying the name of Hoffman to the frame itself. If any name is to be applied to the frame it should be that of the originator.

The decades which have passed since modern beekeeping became possible through the work of Langstroth have seen changes both good and bad. Good leadership has sometimes given way to poor leadership. Beekeeping will attain greater stability only when much less attention is given to fads. And there have been fads since Langstroth's day. When food adulteration was common, most beekeepers produced section comb honey, which could not be adulterated and was therefore a guarantee of purity to the consumer. Many useful practices were then devised, with regard to swarm control, for instance, and the order in which supers should be placed on the hive. That brilliant era closed under the leadership of Heddon and his associates. In order to crowd honey into the supers, they reduced the brood space and thereby brought on no end of evils—poor wintering, excessive swarming, weak colonies, and other sources of heavy loss. From a period of brilliant progress, the comb-honey era quickly declined and was followed by a general recession.

The enactment of the Federal Food and Drugs Act in 1906, and the simultaneous experiments of energetic beekeepers who wished to save the industry, made it possible to embark on an extensive production of extracted honey. The fine art of producing comb honey has now been largely lost, but newer methods enable beekeepers to care for more colonies, to produce much more honey, and to use labor-saving devices. Beekeeping of today is the outgrowth of this change, and there is no prospect of a return to comb-honey production, which may be as well.

In Langstroth's day and for years thereafter all investigations were conducted by practical beekeepers. Serious losses and a desire to promote the industry led to the establishment of state and federal laboratories for attacking problems too complex or too protracted for the apiary, and work on various aspects of bee behavior in these laboratories has clarified numerous questions. The foundation laid by Langstroth was firmly built on a knowledge of the activities of bees, and all the useful contributions of the laboratories have been made on the same basis. The work of the laboratories has in some measure, however, discouraged private investigations, and that is regrettable.

The poorer leadership which has occasionally arisen in beekeeping usually has its origin in human nature. If one has drifted into poor practice, it is natural that he try to rationalize it by seeking to provide proof of the propriety of his methods. Such mental gymnastics do not solve beekeeping or other problems, and it is unfortunate that some of the men who have tried to defend poor practices have received a hearing, to the untold harm of those who followed their leadership. Fads such as top entrances in winter, inadequate protection to bees in cold weather, incorrect placing of supers, all have cost the beekeeper much honey. There are too few Langstroths in present-day beekeeping.

Among beekeepers there is a natural desire to associate with those having similar problems. This has resulted in the formation of numerous associations, yet, unlike similar organizations in several foreign lands, the associations of beekeepers in the United States have remained weak in numbers and in influence. This is probably due not so much to inadequate direction as to the fact that beekeepers have found their own solutions of most of the problems. Coöperative organizations for the sale of honey have been formed and some have failed, perhaps largely because of the individualism of beekeepers.

Any person interested in establishing a library of beekeeping literature will be amazed when he learns of the large number of books that have been written in this field. There are also almost uncounted journals of the past and present. Before the days of Langstroth there were many books, now chiefly interesting as history or as curios. Moses Quinby issued his noteworthy book, *Mysteries of Beekeeping,* in the same month in which Langstroth published the first edition of his classic work, and both books have gone through numerous editions. Other books have since appeared in the United States. Several beekeeping journals have been begun, but most of them have been short-lived. Strangely enough, fewer books about beekeeping are available now than before Langstroth's day. This may be due in part to the emergence of state college bulletins on various phases of beekeeping, but such occasional monographs can hardly replace good books or journals of current interest.

In countries where the German language is spoken there has been tenacious adherence to the Dzierzon type of hive, opening at the rear, although a few persons in those countries who have ventured to use hives of the type devised by Langstroth have met with greater success within the limits of the honey crops there. The entirely human explanation given by other beekeepers in such cases is that the local area of the experiment must be far better for nectar secretion!

Loss from the diseases of bees is being reduced. That is a problem of which Langstroth knew little. The causes of the two important brood diseases are known. European foul brood, a serious menace in the days when colony strength was usually low, is no longer a problem to the good beekeeper. American foul brood is coming under control through rigid official inspection, and the one phase of this reduction which is discouraging is that many beekeepers depend too largely on official inspectors and less on their own alertness. Diseases of adult bees, entirely unknown in the days of Langstroth, do much less damage in North America than in Europe, and this contrast suggests that the proper remedy is the maintenance of colony strength, a practice in which American beekeeping is superior.

Beekeepers look hopefully for a day when the merits of honey will be more widely recognized. Efforts to promote its use as a food and in industry have been sporadic and feeble. Some exaggerated claims have been made, but the merit of honey as a food is self-evident from its flavor. The important next step is obviously to refrain from damaging honey by unwise preparation for market and to control changes which normally occur, such as crystallization. All the honey that this country can produce will sell itself if it is so handled as to conserve its worth.

This tribute to Langstroth's memory appears as the United States enters the second World War, and no prophet is at hand to tell us what effect this crisis will have upon the industry which he founded. In the first World War beekeeping as an industry evolved quickly out of the hobby cultivated by the enthusiasts forming that fine group often called "the beekeeping fraternity." Commercial honey production was then only getting started, and it grew more rapidly in the years of that war than ever before. Since the effects of the war economy wore off some have gone out of the business but others have increased their holdings until today many persons operate hundreds and even thousands of colonies. Skill in colony management seems to have declined in the effort to establish honey production as a business. The organizations of beekeepers have remained weak, the literature of the subject has failed to keep pace with recent changes, and the industry is still largely an individual affair. A few coöperative associations have been formed and a wholesale market for honey developed mainly as effects of that war.

Once more beekeeping is faced with a crisis brought on by a world conflict. If, through thoughtful study of its problems, wise counsels to honey producers, effectual organization of various subsidiary interests, and a general coöperation throughout this branch of agriculture, honey can be produced on such a scale and marketed so effectually as to make its food value a factor of the national economy, then beekeeping may be expected to thrive. Many vocations have had their rise or fall in changing social, economic, or political conditions, and beekeeping now has a rare opportunity to prove its great usefulness to the community. Competition from other sweets is more keen than at any previous time, and those who produce them will doubtless seek to promote their own interests. The beekeeper, in safeguarding his industry, may fairly claim to be serving the community in a peculiar way. The pollination of plants by honeybees is so necessary to the welfare of agricultural areas that the survival and growth of beekeeping is a matter of national moment. Granting that, however, beekeepers cannot escape responsibility for their own problems.

Followers of Langstroth today are not lacking in skill, vision, or courage. If they can combine their efforts in adjusting their affairs to the changing conditions of our times, there will be ground for hope that this branch of agriculture will become an even greater monument to his early leadership and invention.

AMERICA'S MASTER
OF BEE CULTURE
The Life of L. L. Langstroth

I

Parentage and Boyhood

LORENZO LORRAINE LANGSTROTH, the subject of this biography, was an American whose life spanned eighty-five years of the nineteenth century.[1] He is remembered for the beneficent rewards of long experimental study which he gave to the culture of the honeybee. His memory is held in honor for the particular reason that he discovered the principle and conceived the plan of the hive with scientifically spaced and easily interchangeable frames, inventing as he did so that ingenious system of interior construction which makes the modern beehive a perfected home for the bees and at the same time a perfected tool for the keeper. Only by the use of Langstroth's invention has beekeeping become the practicable, remunerative, and enjoyable branch of agriculture that it is today.

He was graduated with distinction at Yale College in 1831, took a course in the divinity school, and was ordained to the Christian ministry, which he practiced,

---

[1] For some of the material of the early chapters the author is indebted to Langstroth's own account, which he wrote in response to A. I. Root's entreaty for an autobiography and which Root published in 1892–93 as a serial in *Gleanings in Bee Culture* under the title "Langstroth's Reminiscences." Fourteen installments appeared, but the autobiography was never accomplished.

with intervals of teaching, until ill-health unfitted him
for a regular charge. While serving as minister of the
South Parish in Andover, Massachusetts, he happened
—or was led—to take up beekeeping as a hobby. It
became his absorbing pursuit. He employed in it a
talent for observation and a patient power of reason-
ing from all that he observed.

He was born on Christmas Day, 1810, at Philadelphia, Pennsylvania, where his grandfather, Thomas
Langstroth, a Yorkshireman, had settled about forty
years before. The circumstances of that settlement have
been kept in memory by family tradition. Thomas
Langstroth, then twenty-two years of age, coming to
America on business for his grandfather, was a passen-
ger on an English ship, the *Glory*, which entered Dela-
ware Bay on an autumn day of 1767. As the ship sailed
up towards her destination in the harbor of Philadelphia, passengers exclaimed at the unexpected beauty
of this new country, observing broad and fertile fields
on either side of the narrowing estuary and beyond
them the primeval forest. No one admired the prospect
more than the young man from the Yorkshire moors.

By the time that the business which had brought him
to America was finished, Thomas Langstroth had given
up the idea of returning to England, although that
meant the breaking of a strong tie. His family for two
centuries or longer had lived in a single parish of the
West Riding, the parish of Horton in Ribblesdale, near
where the River Wharfe rises in a long strath or moun-

tain valley which bears the name of Langstrothdale.
You will find the name on any large-scale map that includes Penyghent, one of the hills of the Pennine Chain,
the backbone of Northern England. But Thomas Langstroth found the Pennsylvania colony so much more to
his liking that he determined to stay there. Philadelphia, the capital, seaport of a rich agricultural region,
was thriving with industry and commerce and seemed
likely to become the metropolis of all the Atlantic sea-
board of North America.

Near Philadelphia, within a few years, Thomas Langstroth set up a mill for the manufacture of fine paper,
one of the first mills of its kind in the colonies. He had
to face difficulties. In those years of the reign of King
George the Third, when the time was getting ripe for
the American Revolution, any colonial manufacture
which competed with the trade of the mother country
was likely to encounter arbitrary discrimination. And
there was a preference for English goods. It was years
before the Langstroth mill ventured to distinguish its
paper with a watermark. But the business did well.
Thomas Langstroth became widely known and respected. He practiced gardening as an avocation, and
the family has not forgotten that his reputation for
skillful horticulture brought him a visit from no less
a person than George Washington.[2] A genial nature,
broad judgment, and sense of fairness won for him,

[2] Philadelphia was the seat of the Federal government during seven of
the eight years of Washington's presidency.

among his Quaker neighbors, the title of Peacemaker. An anecdote has been preserved which illustrates his character. One winter evening, towards the close of his life, after reminding his wife how greatly Providence had prospered them, he showed her some papers which, he said, would be enough to establish his title to a considerable estate in England. But, he said, his kinsfolk there had much greater need of it, and with his wife's consent he would like to burn the papers. She consented, and the papers were thrown into the fire.[3]

Three years after his settlement in Pennsylvania Thomas Langstroth married Anne Youck (anglicized George), whose parents had come from Prussia to the Germantown suburb of Philadelphia. They had twelve children of whom four died in infancy. The fourth son, John George Langstroth, succeeded to his father's business and became the owner of two paper mills. He married, in August of 1808, Rebecca Amelia Dunn, daughter of James and Elizabeth (Lorraine) Dunn of Chestertown, Maryland. The first member of this Dunn family of whom there is record was Robert Dunn, who came from England in 1649, settled on Kent Island in Chesapeake Bay, and afterwards removed to Chestertown on the Eastern Shore.[4] The family's home there was built of bricks brought from England as ballast in vessels

[3] L. L. Langstroth wrote in his Reminiscences that he had this anecdote from the lips of his aged grandmother. He added that he had often sat before the old Franklin stove in which the papers were burned. *Gleanings in Bee Culture*, xx, 761.
[4] The year of Robert Dunn's coming to America and the fact of his settlement in Lord Baltimore's patent might indicate that he was a royalist.

which were to return laden with tobacco from the Dunn plantation. The family took high rank in the social and political life of the Maryland colony. A Dunn was among the founders of Washington College at Chestertown in 1782. Elizabeth Lorraine, Rebecca Dunn's mother, was a granddaughter of Count Louis Lorraine, a Huguenot of noble family who, after the proscription of the Protestants in France by Louis the Fourteenth's revocation of the Edict of Nantes, chose exile and poverty rather than renounce his faith and came with the Huguenot emigration to America.

The death of James Dunn left his wife with considerable wealth, including numerous slaves. She did her best to treat them humanely, even teaching them to read when it was illegal to do so. Before her death she determined to set them free, having, it is said, been convinced of the evil of slavery by the teachings of John Wesley. To the men and women who were able to take care of themselves she gave liberty without restriction. The boys were bound out to service until they were twenty-one years old, the girls until eighteen. Provision was made for the aged and the disabled. The once wealthy widow thus deprived herself of all but a modest competence.

John and Rebecca Langstroth lived in Philadelphia at No. 106 South Front Street, not far from Independence Hall. Eight children were born to them. The second of these children, the eldest son, who was named Lorenzo Lorraine, was in one respect a peculiar boy.

Very early in life, as he has recorded,[5] he began to take "an extraordinary interest in observing the habits of insects." When he was about six years old, so his mother afterwards told him, his teacher reported that although he was doing well in other respects she had had to punish him for spending time in catching flies and shutting them up in paper cages. He remembered crying himself to sleep in a dark closet where the teacher had put him after tearing up a cage and letting out his flies. His narrative continues:

Although my parents were persons of good intelligence, and in comfortable circumstances, they were not at all pleased to see me spend so much time in digging holes in the gravel walk and filling them with crumbs of bread, pieces of meat, and dead flies, to attract the roving ants, so that I might better watch their curious habits. I know that I was once whipped because I had worn holes in my pants by too much kneeling on the gravel walks in my eagerness to learn all that I could about ant life. No books on natural history were given me; but I was considered a foolish boy whose strange notions ought to be severely discouraged. But nothing that could be said or done prevented me from giving to my favorite pursuits much of the time which my school companions spent in play.

He has given a particular account of what he noticed of the habits of the periodical cicada, often but wrongly called the "seventeen-year locust," and other species closely allied to it—some of which could be found every year. The story follows:

I could not have been much over eight years old when these locusts first attracted my attention. Year after year I visited the

Center Square, a public park of Philadelphia, to secure specimens and to study their fascinating transformations. These insects came out of the ground late in the afternoon, and I noticed
that the holes out of which they crept were as smoothly bored
as though made with an auger. As soon as an insect emerged
from its hole it made for a tree or some other object, up which
it would creep to a satisfactory height. If suddenly approached
while in the act of mounting, it would often, " 'possum-like,"
drop to the ground as if dead. After fastening its sharp claws
into some chosen surface it remained motionless for a short
time.

When a larva [6] first leaves its hole in the ground, its body
feels quite hard, but before long it becomes almost as soft as
dough. Now, in its soft state it can no longer crawl; and if prevented, before it became helpless, from getting a firm hold on
some object, it would be quite impossible for it to emerge from
its shell; but, fastened firmly by its claws, it soon began alter-
nately to contract and expand its body, until what at first resembled a little crack on its back opened wider and wider, continually disclosing more and more of the emerging insect, until
at last it raised its head and the larger part of its body from
the shell, being prevented from falling out of it by the lower
part of its abdomen, which was still held in the shell. It then
looked considerably like an Egyptian mummy standing up-
right in its case, with its upper wrappings removed. Now, as
the transition from their hard to their soft condition is a very
short one, it is obvious that these holes, which are often bored
through hard ground, must be made a considerable time before they are wanted, to enable the insect to push quickly
through the little space that is needed to let it out, when its
instincts teach it that the time is at hand for its coming changes.

[6] The periodical cicada undergoes four moults during its larval life. In
the second pupal stage it emerges from the ground, near where it entered
seventeen years before, and seeks a support where it becomes transformed
into the adult. Langstroth, being familiar with the larvae and the motionless pupae of the bee at the time when he wrote this, seems to have thought
that the insects which he had seen crawling from the ground in years long
past were larvae and not pupae.

As soon as it has withdrawn its head, legs, and other parts of its body from the horny shell in which each was separately enclosed, it rests a while until its claws, which at first are too feeble to grasp anything, become strong enough for it to climb out of its shell and cling to the rough surface on which it had fastened itself. Its wings, which are narrow and thick, can now almost be seen to thin out gradually, like a piece of dough over which a roller is continually passing. When they reach their full expansion they remain thus flattened out until they become quite dry, when all of a sudden, by an involuntary motion, they assume the proper position for flight.

The locust lives only a short time as a flying insect, when the female bores holes into the extremities of small twigs, in which she inserts her eggs. The larvae, when hatched, feed upon the twigs until the latter wither and fall to the ground, when they penetrate the earth, to reach the roots of trees, by sucking the juices of which with its sharp, hollow proboscis (as has recently been discovered) one species lives for seventeen years.

I was not over twelve years of age when I made most of my observations upon these locusts; but when I returned to Philadelphia in the fortieth year of my age, it being locust year, I collected quite a number of larvae from the trees in Independence Square, and sat up with my daughter and some of her school companions until after midnight to show them the curious changes just described; but from my boyish recollections I could have described them almost as vividly and accurately as I could after these last observations.[7]

At school he evidently earned good reports, for his father was not slow in making plans to send him to col-

[7] The various broods of the "seventeen-year locust" have been carefully recorded for so many years that it is possible to determine which of them Langstroth observed. The first one that he noticed was doubtless Brood X, which appeared in Philadelphia in 1817, his seventh year. Its latest appearance was in 1936. Brood XIV appeared there in 1821, his eleventh year, and completed its latest cycle in 1940. Brood II made an appearance in 1850, his fortieth year. Brood X (1817) is so abundant about Philadelphia that an observant child could hardly miss seeing it.

lege. In due time he was enrolled in the preparatory school conducted by the University of Pennsylvania. There too he appears to have acquitted himself well. He became notably proficient in Latin. He remembered the principal, the Reverend James Wilbanks, as an accomplished teacher of the classical languages, and as a master who had also the faculty of communicating to some of his pupils a love of the literature. In the course of his study of Vergil young Langstroth committed to memory hundreds of lines of the *Aeneid,* and many of them he never forgot. There is evidence of that accomplishment in a granddaughter's story of her own experience years afterwards.

"When I was studying Vergil," she said, "I used to ask Grandfather to help me with the translation. But, instead of translating, he would begin reciting the Latin. That did not seem very helpful to me, not until I found out that he was getting the meaning directly from the Latin. I thought he must know the whole poem, for he could begin and go on wherever I happened to be at work in it."

Langstroth's reminiscences included another good reason for a clear memory of his Latin teacher. He told it as follows: "Mr. Wilbanks was a disciplinarian after the very straitest sect of the Old School. If I was late, and had no excuse, I always stepped up to his desk, and held out my hand and took my punishment with as much grace as I could. The rod! the rod! this was the universal arbiter, from which there was no appeal. I once pronounced the word 'a-*mi*-cus' as though it were 'am-i-cus.' In a thin voice, so shrill as almost to resemble a squeal (I can almost imagine that I still hear it ringing in my ears) he cried out to me, '*Am*-i-cus, Lorenzo! I'll *am*-i-cus you! That word is a-*mi*-cus!' And down came his rod with such an effective emphasis that I never forgot to say a-*mi*-cus. But although he used the rod so freely, it was only in the way of what he thought his duty, and I never associated his name with any thought of cruelty. He made me a good Latin scholar, and his memory will always be a pleasant recollection."

From the preparatory school Langstroth went to Yale.

II

Student and Tutor at Yale

LANGSTROTH was admitted to the Freshman class at
Yale College in the autumn of 1827, in his seventeenth
year. The town of New Haven, which then shared with
Hartford the honors of capital of the State of Connecticut, had a population of about ten thousand. The main
buildings of the college, facing "the Old Green," were
at the center of town. There were four large halls occupied by students as studies and dormitories, two dining halls, one of them for students of the divinity
school, a chapel, the president's house, and a few
smaller buildings.

Candidates for admission to the Freshman class were
examined in a selection of Cicero's orations, Vergil, and
Sallust; the Greek Testament and "Graeca Minora";
English, Latin, and Greek grammar; Latin prosody and
writing; Arithmetic, and Geography. Instead of the
Greek Testament and "Graeca Minora" the candidate
might choose to be examined in the Greek Reader and
the four Gospels.

The four-year course of study, which was a curricu-
lum or specified fixed course leading to the bachelor's
degree, was substantially as follows:

Freshman Class: Folsom's Livy; Adam's *Roman Antiquities;* Graeca Majora begun; Day's Algebra; Horace and Euclid begun.

Sophomore Class: Horace and Euclid finished; Graeca Majora continued; Day's Mathematics; Plane Trigonometry, Nature and Use of Logarithms, Mensuration of Superfices and Solids, and Isoperimetry; Mensuration of Heights and Distances, and Navigation; Juvenal, in Leverett's edition; Cicero *De oratore* begun; Surveying; Bridge's Conic Sections; Spherical Geometry and Trigonometry; Rhetoric.

Junior Class: Cicero *De oratore* finished; Olmsted's *Natural Philosophy* and *Mechanics;* Tacitus, the *History,* the *Manners of the Germans,* and the *Agricola;* Astronomy; Graeca Majora continued; Hedge's Logic; Tytler's History. At the option of the student: Fluxions; Homer's *Iliad;* Hebrew, French, or Spanish.

Senior Class: Blair's Rhetoric; Stewart's *Philosophy of the Mind;* Paley's *Moral Philosophy;* Paley's *Natural Theology;* Brown's *Philosophy of the Mind;* Greek and Latin; Evidences of Christianity; Say's Political Economy.

In addition to recitations in all the above requirements there were courses of lectures in Languages, Natural Philosophy, Chemistry, Mineralogy, Geology, and Astronomy. A course of lectures on the oration of Demosthenes *On the Crown* was delivered to the members of the Senior class. Specimens of English composition were exhibited daily by one or more of the divisions of the Sophomore and Junior classes. Written translations from Latin authors were submitted by members of the Freshman class, and the lower classes were drilled in Latin composition. The Senior and Junior classes had "forensic disputations" once or twice a week before their instructors. There were regular exercises in declamation before the tutors, before the professor of oratory, and before the faculty and students in the chapel.

Every student was required to attend prayers in the chapel morning and evening of every weekday, and to be present at public worship in the chapel on Sunday unless he had special permission to go to one of the churches in town.

There was little if anything in the prescribed course of study which Langstroth took at Yale College that could have prepared him for such a pursuit as that of his later years—the patient observation and study of the life of the honeybee. Perhaps the nearest approach to what is now the subject of biology was the required reading of William Paley's *Natural Theology, or Evidences of the Existence and Attributes of Deity*, published in 1802, a book which, seeing in nature only evidences of design and purpose in a single act of creation, has long been ignored by naturalist and theologian alike. There seems to be nothing to show that Langstroth's life work was affected by any theory old or new. He owed nothing to Paley's "evidences." He does not appear to have accepted the evolutionary doctrines so hotly debated during the period of his intensive study of the bee or even to have been concerned with them. There is nothing to indicate that the course of his work would have been materially changed if he had adopted the newer opinions. But certainly his method was the new and revolutionary one: for answers to his questions about bees he went to the bees themselves as the only source of sure information. His mature work reveals the curiosity, open mind, and adherence to ascertained fact which are marks of the true investigator. On the other hand his life gives evidence of the sensitive nature and clear insight which keep a religious faith untroubled by shifts of theory.

Apart from his regular studies there was one circumstance of Langstroth's residence at Yale which may well have given him more than a hint of methods which he afterwards made his own in his study of bees and his development of the hive. That circumstance was his friendship with Denison Olmsted, professor of mathematics and natural philosophy. He mentions it significantly in his Reminiscences.[1] He sums up in a brief paragraph the nature of his studious interests at college. Despite his "early passion for investigating insect life," he says, he can not remember, "with the exception of a few trifling observations upon the habits of glowworms," that he took the slightest interest in his old pursuits. The paragraph concludes: "My attention was mainly given to mathematics and *belles-lettres* studies, and I was always among the successful competitors for excelling in English composition. I roomed, in my freshman year [1827–28], at the house of my college guardian, Prof. Denison Olmsted, who had charge of the college meteorological observations, and who in-

[1] *Gleanings in Bee Culture,* xx, 796–7.

spired me with a great fondness for his favorite pursuits."

What those "favorite pursuits" were is a matter of
record.[2] Olmsted, then thirty-six years old, was full of
scientific curiosity, busy with inventive enterprise, and
already making a reputation. He had graduated at Yale,
given a year to special study under Benjamin Silliman,
and gone in 1818 to be professor of chemistry and geology at the University of North Carolina. There he
had brought about and accomplished a survey of the
state's geological and mineralogical resources, the first
survey of its kind in the United States. Called back to
Yale in 1825, he was now extending his studies to phys-
ics, astronomy, and meteorology. At this particular time
he was studying hailstorms. His observations were giving him material for an article on hail formation, published two years afterwards, in which he showed the
electrical theory then held to be incorrect and gave
substantially the explanation, based on dynamics and
thermodynamics of the atmosphere, which is accepted
today. Olmsted as a scientist appears ahead of his time
in some respects. He advocated a laboratory for scientific research and inaugurated laboratory work for
his students. In his own researches he had an eye open
to useful applications. He invented an inexpensive
lubricant, an improved stove, and other devices. Although his work, so far as Langstroth could have heeded

---

2 See, e. g., Alois F. Kovarik's sketch of Olmsted's life in the *Dictionary
of American Biography*.

it, may have had little if anything to do with biology, yet there is at least a possibility that it did set this impressionable pupil of his a memorable example of ingenious observation and practical objective.

Langstroth happened to go to Yale at a time when the students of the college were chafing under a new strictness of discipline effected by President Jeremiah Day. Under his administration the faculty had been made responsible for the government of the students. There were occasional revolts. Langstroth found himself in the thick of one of them before he had been there a year. This is his recollection of it:

In the summer of 1828 there occurred what will ever be famous in the history of Yale College as the great "Bread and Butter Rebellion." The students were all required to board in commons, unless they could procure a physician's certificate that their health required a different diet. The summer was unusually hot. The bread was not always sweet nor the butter fresh, and loud were the complaints against the regular fare. At a meeting of the different classes a resolution was unanimously passed that the students should show their dissatisfaction by absenting themselves one Monday morning from the dining hall. Word had come to our venerable president, Jeremiah Day, of what the students purposed. So after morning prayers, which he usually conducted, he addressed them, in his wonted kind and courteous manner, telling them that, if they had causes of complaint about their fare, they ought in a respectful way to make them known to the faculty, whose interest it certainly was to have them remedied. He closed his appeal by affectionately warning us against any hasty and improper proceedings, which could result only in evil. But our passions were too much inflamed, and we were too much under the influence of those who had planned the original demonstration, to listen to anything our good president could say. So when the bell rang out the summons for breakfast, crowds gathered around the dining hall. None entered; but all, with loud shouts of defiance, expressed what they thought to be a proper sense of their wrongs.

Before dinner the classes met again for consultation, and their leaders now advised that they should decline to take any meals in commons until they had sufficiently expressed their indignation for the kind of food which had been served to them, and had obtained assurances from the faculty that their grievances should be redressed. Thus was inaugurated an absolute rebellion against the constituted authorities.

Before entering college I had promised my parents to obey its laws, and to give them no occasion to regret the sacrifices which they were making in my behalf. I saw that the course which we were now pursuing was a direct violation of that pledge; and without consultation with any one, I determined to retrace my steps and to go into the dining room at the next meal, even if I went alone. At a meeting of our class I announced this determination, saying that we all knew we were violating our matriculation pledges, and that, while I did not profess to be governed by a higher sense of right than others, I did intend to redeem as far as I could the promise which I had made to my parents.

I left the meeting after these remarks, and a committee was appointed to remonstrate with me, and to assure me that, if I persisted in my intentions, I should be treated by the whole class with merited contempt. The hour for dinner arrived, and the students were assembled in unusual numbers, as the report of what I meant to do had become generally known. Yells of execration greeted my appearance, as alone I ascended the steps leading into the dining hall; stones were thrown at me; and one student, more daring than the rest, drew a pistol and threatened to shoot me. Nothing, however, could move me, for I was nerved to such a pitch of determination that I would have submitted to instant death rather than change my purpose.

In the afternoon of that day, my guardian, Prof. Olmsted, who knew nothing of my intentions until all was over, informed me that, by vote of the faculty, I had been excused from entering the hall again, and that my safety, and his duty to my parents, demanded that he should prevent it. I told him that they might kill me, but that I would never yield to them; and when I entered again, quite a number, most of whom I think were professors of religion, were emboldened to enter with me.

The issue of this affair was, that some students were expelled from college, all recitations were suspended, and the students returned to their homes. Only after signing due apologies were they allowed to resume their studies, at the beginning of the next college year. The course which I had taken, although at first so unpopular, in the end made me a host of friends. It was probably the turning point in my life, for my natural disposition often inclined me to yield my own convictions of duty in order to be on the popular side. It would be difficult to tell how much I owe to that "Bread and Butter Rebellion." [3]

Not until his senior year in college did Langstroth become especially interested in religion. Some of his friends there were "professors" of religion and some were not. Of the latter, however, he recorded this opinion: "I believe that I can truly say that we aimed to cherish a high sense of honor and purity, and that our mothers and sisters might have heard without a blush our most private conversations. Since I have mingled freely with men, I am quite persuaded that what I have said of our conversation is very far from being a common occurrence." [4] In the autumn of 1830 a student

[3] Reminiscences, *Gleanings in Bee Culture*, xx, 796–7.
[4] Reminiscences, *Gleanings in Bee Culture*, xx, 832–3.

named Peter Parker came to Yale from Amherst College and was admitted to the Senior class. Within that academic year he began among the students a religious movement which became known in Yale annals as "the great revival." Langstroth, who had been "living without prayer and without reading [his] Bible," was caught up by it.

One day when he was ill he had a visit from Parker, who turned the conversation to the subject of religion. Langstroth was interested, but when he was well again he was disposed to avoid the subject. One afternoon, however, when recitations were over, Parker fell in with him and proposed a walk. Knowing Parker's object, he agreed reluctantly, meaning to get rid of him as soon as he could. But they walked and talked for some time, and when they turned back it was to call upon the Reverend Chauncey Allen Goodrich, the professor of oratory and rhetoric, who had encouraged Parker's missionary activities. Langstroth was the first student won over, but before long there were many others, including, as he recalled, two of his classmates who were afterwards eminent, Noah Porter as president of Yale College and Lyman Hotchkiss Atwater as professor of philosophy at Princeton.

The Class of '31 at Yale had not a few members who earned distinction in later life. Besides Langstroth himself, Porter, and Atwater, there were James Hopkins Adams, Governor of South Carolina in 1854–58; Thomas March Clark, Bishop of the Episcopal Church in Rhode Island for forty-nine years, 1854–1903; William Ingraham Kip, the first Bishop of the Episcopal Church in California, 1853–93; Trusten Polk, Governor of Missouri in 1856–57 and then a United States Senator until 1862; Alpheus Starkey Williams, brigadier general in Sherman's army, Congressman, and United States minister to San Salvador; and the Peter Parker already mentioned, who became the first Protestant medical missionary in China and represented the United States Government there as resident commissioner and minister.

Langstroth's high standing as an undergraduate student was attested by his election to membership in the society of Phi Beta Kappa, then as now a recognition of scholarship and character. The chapter of which he was a member is one of the oldest, the Connecticut Alpha, established in 1780 by charter from the parent chapter, Alpha of Virginia, which had been founded at the College of William and Mary in 1776.

In the spring of 1831, just before his graduation, Langstroth joined the College Congregational Church in New Haven. In the autumn of that year he entered the divinity school at Yale. He planned to support himself and meet the expenses of the course by teaching. His father could not any longer afford to give him an allowance, and he was unwilling or unable to obtain aid from the American Education Society, which the Congregationalists had founded in 1815 for the purpose of helping young men to prepare for the ministry. As a rule that society's funds were used for the most part to assist students before rather than after graduation from college. He found employment as a teacher in one or another school for young women in New Haven; during one year he taught school in a village across the Hudson River from West Point.

In the autumn of 1834 he was appointed tutor in mathematics to the Freshman class at Yale, where he served during the next two years. He succeeded to a remarkable degree in winning the respect of his students. At that time it was no uncommon thing for tutors to be "smoked out" of their rooms or to have their windows broken. Langstroth was free from any such annoyance. A courteous manner, springing from an unfailing kindness and a personal interest in his students, probably earned him the immunity. Their respect and liking for him are illustrated by anecdotes.

One morning in winter, when prayers were held and recitations begun before sunrise, Langstroth, who had been kept awake in the night by illness, overslept and missed an early class appointment. Ordinarily, following an immemorial custom, the class would have waited only a few minutes and would then have dispersed noisily. But on this occasion they sent a committee to the tutor's room to say that they knew his absence to be due to the state of his health and that they were waiting for him. Such an act on the part of freshmen was so unheard of that it made a good deal of talk. When again, a few days later, he overslept and a class delega-

tion waited upon him there was even more talk. Then
he told the class that he was grateful for their kind consideration but that if he ever again failed to meet them
on time they were to feel free to dismiss themselves.

Another incident of that year was related by a member of the same class at a reunion some years afterwards,
when Langstroth was present as a guest of the class. This
member told how the young tutor had caught him
pounding on the door of a room where a recitation was
being held. That was a prank which had been giving
so much annoyance that the college authorities had
publicly threatened to expel any student caught at it.
Instead of reporting the offender, Langstroth had sent
for him, talked with him, and offered to say nothing
about the matter if the boy would give his word of honor
not to do such a thing again. The man who told the
story at the reunion added that he had made the promise
and kept it, and that the experience of such trust by a
man of Langstroth's reputation for strict adherence to
discipline had made a man of him.

Langstroth had been awaiting an opportunity to enter
the ministry, and now it came.

III

Minister and Schoolmaster

EARLY IN 1836, while he was a tutor at Yale College, Langstroth was called to become minister of the South Church in Andover, Massachusetts. He had made himself a candidate for that appointment by accepting the church's invitation to preach before the congregation on two Sundays in the college's winter vacation, and the call came to him almost immediately afterwards. A terse record of it appears in the church's minutes, as entered and signed by the clerk, Amos Abbot, under date of January 7, 1836:

Church meeting in the vestry. The Meeting was opened with prayer by the Moderator. After several preliminary remarks, a motion was made and seconded that we give Mr. Lorenzo L. Langstroth an invitation to settle with us in the work of the Gospel ministry, and the motion passed by a unanimous vote. Chose the Deacons of the Church a Committee to communicate the doings of the Church and invite the concurrence of the Parish. Voted that Deacon Mark Newman be a Committee to communicate the foregoing to Mr. Langstroth.

The church had been without a minister for some time and had heard one candidate after another without coming to a decision. Now that there was agreement the deacons were urgent that Mr. Langstroth accept the call. On the advice of men who had been his teachers in the divinity school he did so. And that may well have been his own inclination, for the South Parish in Andover was one that any young candidate might have been proud of a call to. It had been organized as late as 1710 but had become famous in New England through its association with certain events. Its first pastor, the Reverend Samuel Phillips, who had served it continuously for sixty years, was the father of the Samuel, John, and William, and grandfather of the Samuel and William, who founded and endowed the Phillips academies at Andover and Exeter. Phillips Andover academy and the Andover theological seminary were within the parish bounds. Langstroth accepted the call. He soon had reason, however, to regret his decision. He found the duties too arduous for him. The parish was large in area and he had the pastoral care of about five hundred persons. Parochial labors were new to him and he was unprepared for regular preaching. To make his situation worse, he was subject to a nervous malady which recurrently disabled him for any duty. His ministry at Andover lasted only two years and then he was relieved of it at his own request.

Brief and burdensome though his stay was, his years at Andover were far from unhappy. He made lifelong friends there, and he always afterwards spoke of the parish and the town with gratitude. If he had a regret, it was in recalling that the tenor of some of his sermons had been too austere. He had, in fact, listened to gentle admonitions on account of that. He must have taken them to heart, for when he wrote his reminiscences in 1892–93 he recorded this admission:

For many years I have been painfully sensible that there was too much law and too little gospel in many of the sermons of my Andover ministry. One day a college classmate who was then in the Theological Seminary came to talk with me on this feature of so many of my discourses. He had been talking with one of my oldest deacons, who had sorrowfully recognized this great defect in the sermons of his beloved pastor. I was not offended by the plain talk, but I could not then be made to feel that it was just.

It has always seemed to me, that to no man so much as to Henry Ward Beecher was owing the great change which has taken place in our preaching. When the great Jonathan Edwards was preaching from the text, "Sinners in the hands of an angry God," we are told that one of the preachers who was in the pulpit with him pulled his garment, and said, in a low voice of remonstrance, "O Mr. Edwards! remember that God is a merciful God."

It is written, "Like as a father pitieth his children, so the Lord pitieth them that fear him. For he knoweth our frame; he remembereth that we are dust." These words seem to me to open a window into the very heart of God, so that every father and mother may know from experience how God feels toward his erring children. Is there one child more easily tempted than the others, more ready to fall? Does the parent's heart go out against that child? Does he know it, in all its weaknesses, only to lay more aggravating burdens upon it? No! a thousand times no! Let us never forget how beautifully the Savior illustrates God's love and pity in the parable of the prodigal son.

I once preached from the text, "Not knowing that the good-

ness of God leadeth thee to repentance." I have no doubt that
the tenor of this discourse was molded largely by my recollection of a discourse preached so often in revivals of religion by
Dr. Nathaniel Taylor.[1] The good old deacon, in speaking of
this sermon, could hardly find words strong enough to express
his delight as he opened his whole heart to me and tried to
make me see how much more I could do to lead sinners to
Christ by preaching oftener on the love and goodness of the
heavenly Father.

At Andover Langstroth met the Reverend Lyman
Beecher, who gave him reason to remember the meeting. It was in the early summer of 1836, when the new
minister, hardly settled yet, was living in a boarding
house. Beecher was then pastor of the Second Presbyterian Church of Cincinnati as well as president of Lane
Theological Seminary. He came on a tour of the Eastern
states for the purpose of interesting young seminarians
in the religious needs of "the Great West." With him
came the Reverend Thomas Brainerd, who had graduated at Andover Seminary five years before and had now
for three years been his associate in the Cincinnati
church. In the two-year interval Brainerd had lived in
Philadelphia and there had married Langstroth's cousin,
Sarah Langstroth. The great preacher's visit was an
event in the seminary town, not least to the young minister of South Church, who had invited him to preach
in his meeting house for the sake of the hundreds who

[1] Dwight professor of didactic theology in Yale divinity school from its
organization in 1822 till the year of his death, 1858. He is reputed to have
been one of the most powerful revival preachers of his day. *Dictionary of
American Biography*, article by C. A. Dinsmore.

would hope to hear him. Beecher had accepted, and the whole town knew that he would preach at an evening meeting. The two visitors were put up at Deacon Amos Blanchard's house, where the young minister was boarding. In the afternoon Beecher asked Langstroth for the use of his study, pen and paper, and freedom from interruption. He was secluded, except for a hasty supper, until Langstroth knocked at the door to say that it was time for meeting. Beecher told him to go ahead with Brainerd and promised to follow in time for the sermon. The preacher entered a packed meeting house and mounted to the pulpit during the "sermon hymn." His text was from Matthew, sixth chapter, thirty-third verse, "Seek ye first the kingdom of God, and his righteousness." Langstroth remembered it as "a grand sermon, crowded with thought, severe in logic, and fairly blazing with his wonderful illustrations." When they returned home Beecher said to him, in effect: "Young man, you know the circumstances in which I wrote the outline of that sermon. But don't trust to what you can do offhand. Take time for study and preparation. This afternoon I was boiling down the studies of years." The manuscript of the sermon outline, a single sheet of paper, was left behind. Langstroth obtained it, kept it till late in his life, and then gave it to the library of Wabash College.

In August, 1836, Langstroth married Anne Tucker of New Haven. She was the second daughter of Mrs. Harriet Tucker, widow of the Reverend James Tucker.

Since her husband's death Mrs. Tucker had conducted a school for girls in New Haven and her daughters had assisted her in its management. Hers was one of the schools where he had taught while he was a divinity student, and it was there that he had made Anne Tucker's acquaintance. According to his own account, their interest in each other ripened over a problem in algebra which she brought to him. Her motive may have been mischievous, for she had been told that the problem was going the rounds of local mathematicians as a trick that was actually insoluble, but young Langstroth retorted by finding a solution. Their marriage was a happy one. She lived until January of 1873. Her husband, who outlived her more than twenty years, wrote of her, "What the wife of Huber was to him in his blindness my dear wife was to me."

The young couple set up housekeeping in Andover under difficulties. Those were hard times for the whole country. A speculative boom was about to collapse in the economic crisis of 1837. Commodity prices were high and the cost of living was far out of scale with fixed incomes. About this time, too, Langstroth's father died and left him responsible for the support of his mother and a sister. His salary, $900 a year, which would then have sufficed for the needs of a family under normal conditions, was hardly enough to make ends meet. The birth of a child—his son, James Tucker Langstroth—in July, 1837, added to the number of his dependents. But when he was beginning to fear that he must fall into debt a piece of good fortune came to him which added materially to his income and brought him relief. He was engaged as private tutor to a young man from Boston who had got into scholastic trouble at college. He owed this windfall to Denison Olmsted, his former guardian at Yale. The boy's father came to him bearing a letter of introduction from Professor Olmsted. The father, as Olmsted knew, was able and willing to pay well for the private instruction and was taking pains to find a competent tutor. This incident, indicative of a continuing friendship, may perhaps be regarded as reinforcing the conjecture—advanced in the preceding chapter—that the example of Olmsted the eager investigator and inventor was not lost on Langstroth the student. The evidence of his friend's regard must have been all the more gratifying to Langstroth for the reason that Olmsted was by now a famous man, widely known for a series of papers in the *American Journal of Science and Arts* concerning the great meteoric shower of November, 1833.

A serious illness in 1838 brought Langstroth to the decision that he must resign the ministry of South Church. He had suffered intermittently for years from a mysterious complaint which he could only describe as a "head trouble." It was painful and enfeebling and was commonly attended with a depression of spirits amounting sometimes to melancholia. The doctors that he consulted could not diagnose it or hit upon a remedy. It seemed to be alleviated by outdoor exercise, and he had been advised to spend as much time as he could in the open air. While he lived in New Haven he had followed this advice by taking long walks out of town. Under the strain of his new responsibilities and duties as minister in Andover his health had failed and now he broke down. The collapse involved even a temporary loss of his voice. When he was convalescent he wrote a letter submitting his resignation. The church's action upon it was taken at a meeting held on July 11 and was recorded by the clerk, Amos Abbot, as follows:

Whereas, the Rev. L. L. Langstroth by a communication dated the 7th inst. has signified his request that the Pastoral Relation existing between [him and] this Church & Society may be dissolved on account of his health and strength not being sufficient to enable him to discharge the arduous duties of his office,

Voted, that for the reasons set forth in his communication his request be granted.

Voted, that the Deacons be a Committee to unite with the Rev. L. L. Langstroth in calling a mutual Council to dissolve the present connection at such time as he & the Society shall mutually agree to be expedient, or as either party may request.

Months passed before the council met to give effect to the resignation. Meanwhile Langstroth served the church and the parish as well as an ailing man could, supplying the pulpit or giving place to candidates, but doing little pastoral work. Nevertheless his salary was paid in full. Finally the action of the council was recorded by the church, with the signatures of Ralph

Emerson, Moderator, and Samuel Jackson, Scribe, as follows:

Pursuant to letters missive from the South Church in Andover, an ecclesiastical council was convened at the house of Rev. Lorenzo L. Langstroth, March 30th, 1839, to consider the expediency of dissolving the Pastoral Relation to said Church. . . . It appearing to said Council that the Rev. L. L. Langstroth had resigned his pastoral charge on account of his health being inadequate to the discharge of the duties of his office, & that the Church and Society had accepted his resignation; & the Council having attended to the statement made by him of reason for dissolving his Pastoral Relation & to the proceedings of the Church & Society in relation thereto, it was voted unanimously that it is expedient that the Pastoral Relation of the Rev. L. L. Langstroth to the Chh. of Andover be dissolved, & it is hereby dissolved.

The Council takes pleasure in bearing testimony to the piety and ability of the Rev. L. L. Langstroth & they sincerely regret his removal from among them; but believing that he is called by the great Head of the Chh. to another sphere, they commend him to the Chhs. of our Lord Jesus Christ wherever the providence of God may call him, as a brother beloved, and as an able and faithful minister of the New Testament. . . .

Langstroth looked about for employment as teacher and found it in Andover, where he became the principal of Abbot Academy. Another failure of health obliged him to resign, as the Academy's records show, "after six months of able and satisfactory service." Then, in the spring of 1840, he was appointed principal of a high school for girls at Greenfield, Massachusetts, and he removed with his family to that town.[2] There, while conducting the school, he supplied the pulpit of the Second Congregational Church for almost two years. He was still hoping to devote his life to the ministry, and when this church invited him to take permanent charge he accepted and was installed on December 20, 1843. He remained there as pastor until 1848, when he resigned, again on account of ill-health. Again he had recourse to teaching. He went to Philadelphia and there opened a school for young women.

Among the letters of introduction which he took with him was one written by Rufus Choate, under date of Boston, July 20, 1848. It was addressed to Joseph R. Ingersoll, son and grandson of eminent Philadelphians and himself a man of repute in the community. It read as follows:

I have ventured to present to you the bearer, Rev. Mr. Langstroth, from a conviction that he is a person of great worth and quite deserving of the kindness of the educated and important portion of Philadelphia. He is a clergyman of high character who is induced of circumstances to devote himself to instruction. In this he has been eminently successful, and I cannot feel a doubt that he will prove so in Philadelphia.

Langstroth took a house at Chestnut and Schuylkill Streets. A new chapter of his life had opened. He had begun to keep bees.

[2] This school occupied a large building, now called the Hollister house, which had been erected in 1796 for William Coleman, afterwards the first editor of *The Evening Post* in New York. The builder was Asher Benjamin (1773-1845), a native of Greenfield, whose *Country Builder's Assistant* (1797) and *American Builder's Companion* (1806, with Daniel Reynard) gave the "late colonial" style a great vogue in the Northeastern states.

IV

Observant Keeper of Bees

LANGSTROTH began keeping bees during his ministry at Andover. His interest in the pursuit was awakened suddenly by a curious incident—curious in the light of the revolution which that very interest of his afterwards brought about in the methods of the apiary. He saw some honey in a "glass globe," a contrivance now obsolete by reason of his invention of the movable frame, and so long obsolete that it now needs explanation.

Extracted honey was little valued and seldom sold in those days. Buyers wanted their honey in the comb. But, until the invention of the movable frame, combs built in the hive had to be cut from their attachments to the walls. Dripping with honey, they were unfit for conveyance or display. As a means of obtaining perfect combs which could be handled and marketed, some beekeepers used glass tumblers, often of globular shape, which they placed upside down in upper boxes ("supers"), directly over holes cut to admit bees from the hive with surplus honey for storage. Such a transparent receptacle, easy to handle and displaying to the full a sealed comb of clear honey, had become a "fancy" article of sale.

What happened at Andover on this occasion is thus described in Langstroth's own words:

I have already said that, notwithstanding my early passion for studying insect life, I can not remember, with a single trifling exception, that I took any special interest in such matters during my college life. In the summer of 1838 the sight of a large glass globe, on the parlor table of a friend, filled with beautiful honey in the comb, led me to visit his bees, kept in an attic chamber; and in a moment the enthusiasm of my boyish days seemed, like a pent-up fire, to burst out into full flame. Before I went home I bought two stocks of bees in common box hives, and thus my apiarian career began.[1]

He goes on to tell how slow he was in learning, how little help he got from books or from experienced beekeepers in his neighborhood, and how he came to rely upon his own observation for increase of knowledge:

With the exception of a small book [now known to have been a work by Jerome V. C. Smith, published at Boston in 1831], the author of which seemed to doubt the existence of such a thing as a queen bee, and my schoolboy's Virgil, I knew absolutely nothing of the vast literature of beekeeping, and of course my progress was very slow. In the end I was undoubtedly a great gainer by this ignorance of books, as what came under my own observation was so carefully studied as to become much more my own.

Almost the very first thing that I bought, when I removed from Andover to Greenfield, was a stock of bees in a [section of] hollow log. Increasing gradually the number of my colonies, I learned, by diligent inquiries of the best beekeepers in my vicinity, all that they could teach. But this was not much, as none of them knew enough to drive bees out of their hives, nor

[1] Reminiscences, *Gleanings in Bee Culture*, xxi, 80–81.

used smoke to facilitate their operations, so that I was indeed groping almost in the dark.

About this time I was fortunate enough to get two valuable works—the "Letters" of the immortal Huber, and the second edition of Bevan's Treatise on the Honey-bee, London, 1838. These works made known to me the facts accumulated for thousands of years by the great masters who had so profoundly studied the habits of bees. I soon became the happy owner of an improved Huber hive, and several bar hives, all made according to Bevan's directions. Ignorant of the futile attempts of Morlot' and other German and French beekeepers to make a practical hive for the common beekeeper, out of the Huber hive, I experimented on that line with no better results until I was content to use the Huber hive merely for purposes of observation.[2] The only improvement which thus far I was able to make upon the hives of others was in giving them greater protection against the extremes of heat and cold and the sudden changes of temperature so eminently characteristic of our climate.

The more he read the more reason he had to know that the way of observation and experiment was indeed the only way by which the hive could be made a much better implement of practical beekeeping. The books left too many of his questions unanswered. Practice seemed to have changed but little during the centuries since Aristotle or Varro or Vergil or Columella had recorded ancient methods for the care of bees.

In recent centuries, or since the invention of the microscope, men had come to know the anatomy and

[2] The hive which Huber had devised has been called a "leaf" hive because it could be opened, like a book standing on end, to expose the surface of a comb. It was planned only for ease of observation and has never been adapted to use in the apiary.

physiology of the honeybee and had thus begun to learn what was going on within the hive. The Dutch engraver and naturalist Jan Swammerdam had first, in the course of a short and laborious life (1637–80), lighted up that mystery with his discovery of the ovaries and the oviduct, proving the sex of the queen, hitherto regarded as a king. The French scientist René de Réaumur (1683–1757) had discovered other truths about the honeybee. The Swiss François Huber (1750–1831), blind from youth, but inspired by his reading of Réaumur and patiently assisted by his wife and an intelligent servant, François Burnens, had uncovered many more secrets of the hive. A German-speaking Pole of Silesia, Jan (or Johann) Dzierzon (1811–1906), was extending the knowledge and advancing the practice of bee culture. But, after all, little improvement had been made in ancient methods of hive construction or of management in the apiary. What had been needed all along was a hive so built as to let the bees live and work in their own way, to give their keeper ready access to the interior for inspection and care, and to permit the removal of surplus combs without waste of honey, disturbance of the colony, or injury of worker-bees.

One step in advance had been the device of bars or slats for the bees to suspend separate combs from. Another had been the use of "wings" to keep the combs separated. Hives of recent introduction, known by the name of the inventor as the Munn, the Prokopovich, or the Debeauvoys, had failed in practice. A hive invented by Dzierzon—publicly described in 1848 after he had tested it for ten years—had comb bars and a door at the back. By means of ingenious management he made it practicable and it came into general use in the German-speaking countries. Langstroth, recalling long afterwards his own early struggles with the bar hive, wrote this summary of its history:

The principle of having the bees suspend each comb from a separate bar, by which Dzierzon accomplished such great practical results, had, indeed, been known before his time. In 1675 (see "A Journey into Greece," by George Wheler, page 411) the beekeepers of Mount Hymettus used it in a rude form, in making artificial swarms; and in 1790 the Abbé della Rocca shows in Plate No. 3 of the third volume of his work on bees, that he used bars with "wings" similar to those used by the Baron [August] von Berlepsch [1818–77], to keep the combs at suitable distances apart. In the sixth volume of Hamet's bee journal, *L'Apiculteur*, p. 146, may also be found Della Rocca's description of his hives, made, as he said, "after the method of the ancient Greeks." On p. 147 of the same journal there is a cut and description of a hive with two tiers of movable slats, and with side-opening doors, invented by Buzairies. A description of this hive may be found in "Mémoires de l'Académie de l'Industrie Française de 1832"; but the inventor says that he made it known to the Society of Natural History in 1828—some seven years before Dzierzon began to keep bees. These proofs that hives with slats, and even with side-opening doors, were used and described prior to Dzierzon's writings, are not given with any intention to detract from his merits, for there is no proof or even probability that he knew of them, but it is very certain that, until by his great skill he had made such hives a success, they had contributed scarcely anything of value to practical apiculture.[3]

3 Reminiscences, *Gleanings in Bee Culture*, xxi, 116–18.

Langstroth was making practical and experimental use of the Bevan hive. It was a bar hive, perhaps as good as any that he could have obtained, but it had the grave defect of its kind: although each bar held a separate comb, the combs when built extended from wall to wall of the hive and could be removed only by cutting them from their attachments. This imperfection became for him a subject of intense study. It was absorbing his attention at the moment, several years afterwards, when he conceived the plan of the movable frame. During his eight years at Greenfield he had little leisure from school or parish duties. When he removed to Philadelphia in 1848 he took a house at the junction of Chestnut and Schuylkill Streets. It had a second-story veranda and several spare attic rooms, and therein he established a city apiary. The next year he set up his apiary in West Philadelphia, about two miles away. There he increased his production of honey and had some success in making it a source of income. He was using hives six inches deep and about eighteen inches square, which gave him a large area for his upper boxes. Specimens of comb honey in glass which he exhibited in 1851 received first award from the Philadelphia Horticultural Society.

Beekeeping was diverting him more and more. The hive held baffling but fascinating problems. As he pondered them there was one thing that he could distinguish clearly, the need of contriving for himself a readier means of access—readier than any hive then afforded

—to the populous interior, if only for the purpose of observation. But not merely for observation: his Journal proves that, as will presently be shown. He knew that a hive freely open to inspection and with freely movable parts was the all-important requisite for practical bee-keeping and had been the goal of inventors before him. As yet he could hardly have envisaged the movable frame of movable sections, each section containing a separate and perfect comb of honey, which eventually became the product of his next experiment.

He attacked his immediate problem, how to contrive a hive cover that the beekeeper could lift easily at any time. What caused the beekeeper's difficulty in lifting the cover? Propolis. Propolis, as Webster defines it, is "a brownish resinous material, of waxy consistency, collected by bees from the buds of trees and used as a cement." Bees use it for stopping crevices in the hive. When packed, as they pack it, between nearly contiguous surfaces of wood, propolis hardens into a strong glue. In plain English it is "bee glue" to many a handler of hives. The covers of the hives that Langstroth was working with became glued so fast that it was hard to pry them off, and when they were replaced the bees glued them tight again. In order to overcome that difficulty he made a practical experiment. It was astonishingly successful. It not only eased the difficulty of removing the cover but led him to a further discovery of the utmost importance. He learned that the builders within the hive had a structural rule which no observer before him had known or surmised. He hit upon the basic principle of the movable-frame hive, the "Open sesame!" to economic beekeeping. He discovered the *bee space*. The story of that experiment and consequent discovery has become a chapter in the history of bee-keeping and deserves a chapter here.

V

Discovery and Invention

LANGSTROTH was trying to make practical use of the Bevan hive, in which the combs hung from individual bars or slats. The ends of the bars rested in rabbets cut in the top of the front and rear walls. The rabbets were planned to be just deep enough to bring the cover to a snug fit on top of the bars. But, however snugly the cover might be made to fit, there were always crevices which the bees filled with propolis, gluing the cover to the bars as well as the walls. He used a simple means of separation between the cover and the bars in order to keep them from getting stuck together. He cut the rabbets deeper, lowering the bars by about three-eighths of an inch.

It may have been only by accident that he made just that much separation. Certainly when he made it he had no thought of doing more than trying to correct a single defect. But months afterwards, when he had noted and considered the fact that the bees were leaving open that shallow space at the top of the hive, and when he was pondering another problem, how to get the combs out uncut, his mind leaped to a solution. Then at last he made the most of his new knowledge: he devised a method of extending the shallow space around the walls of the hive and of converting the bars into frames as a means of getting the combs of honey out in marketable condition and with the least possible disturbance of the workers. All these months his mind was taking account of this strange fact—a shallow space which he had made and the bees had respected—a fact which he could not remember having ever heard of. What he had done, as we know now, was to discover the structural principle of the *bee space,* or, as it might be called, the hive corridor, a space of not more than three-eighths of an inch, which bees leave open for their passage. A space too narrow for their easy passage they seal with propolis; a much wider one they utilize for comb.

With the shallow space at the top he found that he had made a marked improvement. In his own words the change had these merits:

[It] not only facilitated very much the removing and replacing of the cover on which the surplus honey receptacles rested, but gave a shallow chamber from which the heat and odor of the hive could ascend freely into the supers, besides admitting the bees to them in the easiest possible manner.

This improved hive had also a bottom board of my own invention which could be opened or shut, even in the most crowded stocks, without crushing a single bee, and which, as the hive stood upon legs, permitted the attachments of the combs to the front and rear walls of the hive to be severed from below as well as from above.

. . . Having by the use of my improved bar hive secured a more perfect control of the combs than I could find in any other hive, I began to see that artificial swarming might be made much more successful than by Huber's method; for about this time I discovered that bees without a queen, if they build comb at all, make it of drone size; a fact unknown to Huber, and fatal to any practical success in artificial increase by his methods.[1]

Throughout the season of 1851 he used hives with the bee space above the bars, meanwhile pondering the problem of how to get combs out without cutting. Only when it was too late for practical experiment that year did it suddenly occur to him that the bee space held the solution of that problem and not of that alone but of all the beekeeper's problems of manipulation and control. Then he envisaged clearly and fully the movable-frame hive, with bee spaces separating the frames from each other and from all the hive walls.

The bee space, that principle of construction which Langstroth discovered and applied, thus became the modern hive's distinctive feature. The hive is a box enclosing parallel frames each of which is to contain a separate comb. Each frame is hung by its top bar, the ends of which are extended and fitted into rabbets cut in the tops of the front and rear walls. The rabbets are cut to provide a clearance of about a quarter of an inch (the bee space) between the tops of the frames and the hive cover. The same clearance is made between frames and walls, between frames and bottom board, and between frames themselves, so that each frame has a bee space all around it except only at the two points of suspension.

1 Reminiscences, *Gleanings in Bee Culture*, xxi, 80–81, 116–18.

The space beneath is provided by a coaming, a quarter of an inch high, along the edges of the removable bottom board on which the hive walls rest. This permits one hive, without a bottom board, to be set on top of another, both under a single cover, with only the bee space separating the two tiers of frames.

It was not without excitement that Langstroth grasped the idea of this plan of construction. His account of the event follows:

In the fall of 1851 I had nearly completed my application for a patent upon my improved bar hive. It will, no doubt, appear very strange to persons not familiar with the ordinary progress of inventions, that the shallow space between the tops of the bars and the board on which the receptacles for surplus honey rested, [the principle] which I proposed to make one of the leading features in my patent, did not at once itself suggest to me that uprights might be fastened to the bars, so as to give the same bee space between [them and] the front and rear walls of the hive, and so change the slats [or bars] into movable frames. But I used the shallow space above the bars for a whole season without ever connecting the two ideas; and then, only when it was too late to make any use of it in the apiary for that year, did the simple idea of the movable frames present itself to my mind.

Returning late in the afternoon [of October 30, 1851] from the apiary which I had established some two miles from my city home, and pondering, as I had so often done before, how I could get rid of the disagreeable necessity of cutting the attachments of the combs from the walls of the hives, and rejecting, for obvious reasons, the plan of uprights, close fitting (or nearly so) to these walls, the almost self-evident idea of using the same bee space as in the shallow chamber came into my mind, and in a moment the suspended movable frames, kept at suitable distances from each other and the case containing them, came into being. Seeing by intuition, as it were, the end from the beginning, I could scarcely refrain from shouting out my "Eureka!" in the open streets.

At that time there was visiting me my college classmate and dear friend, the late Rev. E. D. Sanders, who afterward founded the Presbyterian Hospital in Philadelphia, and who had taken that season a lively interest in my apicultural experiments. Full of enthusiasm, we discussed, until a late hour, the results which both of us thought must come from using movable frames instead of bars.[2]

Even after his visitor had retired Langstroth could not rest until he had opened his Journal and put on paper a record of the whole new plan. He not only wrote a description and drew sketches to illustrate it, but he also went into details of operation. The entry is dated October 30, 1851. Following are extracts:

If the slats are made so that *a* and *b* are about three-eighths of an inch from the sides of the hive, the whole comb may be taken out without at all disturbing it by cutting. [See Fig. 1, p. 77.]

If the apiarian has some clean worker comb he should fasten it at *1, 2* [see Fig. 2, p. 77]; if not, he should draw a thin line of wax across the center of the bar. If every other bar can be furnished with a comb-guide, it will answer every practical purpose. *a b* should be an inch wide, ¼ inch thick; *1, 2, 3, c, d,* ¾ inch wide. If *a b* is not over 12 inches long, *2* may be dispensed with; *c, d,* about ⅜ of an inch from the bottom board. By the use of such a compound bar, the removal of bars with comb, brood, or honey, can easily be effected. With the ordinary bar, the work of removal is always difficult and often impossible,

2 Reminiscences, *Gleanings in Bee Culture,* xxi, 116–18.

and this is the reason why hives with bars, notwithstanding all their theoretical advantages, have been so little used. It is very obvious that the box or boxes for the storage of surplus honey may be furnished with these bars. . . .

This will be a most excellent way of taking the honey where honey in the comb can be retailed to advantage in a market near to the apiary. The bars in the box for surplus honey should be made the same size as those in the main hive, and thus they may be used for feeding destitute colonies and for artificial swarming. . . .

The use of this bar will, I am persuaded, give a new impetus to the easy and profitable management of bees, and render the making of artificial swarms an easy operation. By the very great ease with which the bars with their combs may be removed, a command over the whole proceedings of the bees is obtained that is truly wonderful. If a hive is infested with the larvae of the bee moth, all the combs may be examined and cleansed in a short time. To one unaccustomed to their scientific management, it would appear to be a very formidable undertaking to remove a bar with its comb full of bees. The timid or inexperienced may wear a bee dress, or resort to a little smoke.

The removing of the queen by means of these bars is very easily accomplished, and this and all other operations may be performed without injuring a single bee, thus preserving the apiary from constant irritation, and keeping the bees always peaceable. It is obvious that the movable frames (I now call them by the better name) may be adapted to almost any hive, and that they will be of the greatest practical benefit.

The compound bars, or movable frames, may be properly suspended and yet the outside cover be movable, with boxes, glass, etc., so as to show the frames without any cover. Such an arrangement will be curious and might perhaps be attended with some practical results. The moth in such a hive would stand but little chance. The tumbler board might form a fixed part of the cover, and the surplus honey might all be taken away at once. A very cheap hive on this plan might be made; it

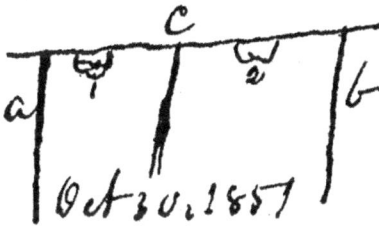

FIG. 1.

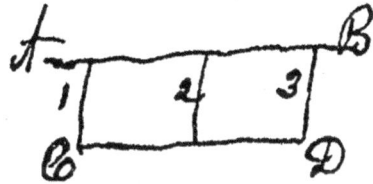

FIG. 2.

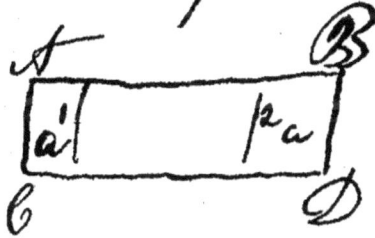

FIG. 3.

Langstroth's diagrams illustrating his record of his invention of the spaced frame (Journal, 1851).

would not, however, be easy to replace the outside cover without killing some of the bees.

The following entry appears in Langstroth's Journal under date of November 25, 1851:

I have this day brought my conception of a new plan of hive to what promises great practical results. My open hive offers the greatest advantages to three classes of persons:

1. The scientific apiarian whose great object is to investigate the habits of the bee. He has a hive where in a few minutes every comb may be examined. Huber's hive, to say nothing of its cost, cannot be used without injuring the bees, and the work of removing frames for inspection is tedious and difficult. The observation hive is expensive, and a very difficult hive in which to preserve the bees over winter. In my hive, he may with the greatest ease perform all the various experiments to arrive at a more accurate knowledge of the habits of the honeybee.

2. The practical apiarian who wishes for profit to manage his bees on an improved and scientific system finds in the use of this hive the means of making artificial swarms, rearing queens, supplying destitute hives with honey or brood, obtaining honey for sale in the market for immediate consumption, or in boxes or tumblers, just as may be most profitable; protecting his hives against the moth, and in short, of performing any operation that may be desired in the spring.

3. The hive will suit the farmer who manages in the old-fashioned way, as he can get honey at any time.

Entries in the Journal for the next two days show that even though Langstroth had brought the idea of his hive to a stage which promised "great practical results," his mind was still busy with the problem. These entries indicate the steps by which he was proceeding:

November 26, 1851. *Cheap* Hive. Movable frames cheaply supported. Shape like figure [see Fig. 3, p. 77]. Cover rough box. Dividers to separate spaces a_1, a_2 from the rest of the hive in winter. Honey to be taken on the bars, bars to work in a_1, a_2 in summer. Such a hive may have tumblers or boxes on top. It may be made cheap and will give the apiarian a perfect control over his bees. A little smoke keeps the bees peaceful, drives them from the comb [which is] to be taken away or [put] into another hive. The webs that worms might spin in the corners of the movable covers easily destroyed, hives examined, cleaned, etc., at all times with the greatest ease. A hive with a movable cover and bars might in a few minutes be taken to pieces, put together again, and all manner of experiments performed upon bees by its aid. The exhibition of such a hive at agricultural exhibits, etc., by an experienced operator would do much to give it currency.

November 27. If one side of the hive cover was movable, the cover might be slid off without so much risk of annoying the bees. The top might be movable like bars put in a rabbet and the hive thus constructed would be cheap and have many advantages. Movable bars and frames the very foundation of all easy management. If there are posts at A B C D the sides of A B C and D may be doubled, filled in with sawdust, or some good nonconductor dividers may be put in in the winter. The entrances may be made by an alighting board under the bottom board which would stand on legs. The surplus honey might be taken from the ends of the main hive on bars or from boxes at the top.

It is evident from these notes that before Langstroth ever actually used a hive with movable frames he had foreseen what came later to be called the "long-idea" hive—one with many frames on a single level, space for the storage of surplus honey at the ends of the hive in-

stead of above, divisible covers, and winter insulation—
including features which some later workers were to
claim as discoveries of their own.

There are entries of the same winter and of the following spring and summer which show that he was
studying the problem of the hive intently and that he
was scrutinizing his ideas carefully until he could put
them to practical experiment in the apiary—always his
crucial test.

The foregoing extracts from his Journal prove that
Langstroth had fully developed his idea of the movable-
frame hive before the end of 1851. They show also that
he knew his invention to be of the utmost importance
for the future of practical beekeeping. About this time
he took certain steps which indicate that he meant to
devote his energies to the perfecting of the new hive and
its introduction into general use. Early in 1852 he applied for a patent, which was granted before the year
was out. He sold the good will of his school in order to
give up teaching. During the next summer he evidently
meant to give all his time to experiment with his greatly
improved hive. His own narrative of the beginning of
that work follows:

Early in the spring of 1852 I moved all my bees to my new
apiary, adding to them a large number of stocks in common
box hives, which were afterward transferred to the movable-
frame hives. This apiary was under the charge of Henry Bour-
quin, a very skillful cabinetmaker, and an enthusiastic lover of
bees. He made the pieces necessary to change my bars into
movable frames; and on the first day warm enough for bees to

fly, the side attachments of the combs to the front and rear walls of the hive were cut, the bees shaken off the combs, and the uprights and bottom strips nailed in place; and so the bar hive became at once a movable-frame hive, in full possession of a stock of bees.

Imagine me so absorbed in manipulating these frames, with the bees upon them—removing from the hive and replacing them—shaking the bees from them, and changing their relative positions, etc., as not to notice the presence of an old beekeeper, nor even to hear him, until he fairly shouted out, "Friend Lorenzo, you are so taken up with your new hive that you seem unable to hear me, or to see anything else. No doubt you think you have made a great invention; but I say you have made no invention"; and then, repeating the words, "you have made no invention," several times to my great astonishment, he wound up by saying, "Friend Lorenzo, you have made no *invention* at all, but, rather, a *perfect revolution* in beekeeping! You have got what I have so long wished for—that control of the combs of a hive, by which you can at any time know the condition of your bees; and, if anything is wrong, be able to apply the proper remedy."

That same summer I had over one hundred movable-frame hives made, some of which were sold with the right to use the patent whenever it should issue; but by far the larger number were publicly used in my own apiary in West Philadelphia.[8]

His affairs did not work out as planned. Before the summer was over he suffered an attack of his recurrent nervous malady and was so prostrated that he could not operate or even supervise his apiary and was obliged to sell it. By November he had obtained his patent and had recovered his equanimity. Then he resolved to leave Philadelphia and go back to Greenfield, Massachusetts,

[8] Reminiscences, *Gleanings in Bee Culture,* xxi, 160–61.

where he hoped to persuade friends to furnish capital for introducing the improved hive. Meanwhile, however, another chain of events led him to undertake a different task. He was induced to write a book, a manual for beekeepers, explaining his methods of management. That book, *Langstroth on the Hive and the Honey-Bee*, and a series of writings of which it was the beginning, eventually did almost as much as his improvement of the hive to make his name honored in the annals of beekeeping. Those writings become the subject of another chapter.

V I

Author and Journalist

§ 1. A MANUAL FOR BEEKEEPERS

A SUDDEN stimulus set Langstroth to work, late in 1852, at the writing of a manual for beekeepers. The impulse came from a new friend, Samuel Wagner of York, Pennsylvania, who was himself an advanced student of bee culture. A chance occurrence made the two men known to each other. Friendship followed upon their discovery that each of them had expert knowledge which in various respects supplemented the other's. The friendship ripened quickly into a productive fellowship. Out of it came Langstroth's book. Eventually it led to the commencement of apicultural journalism in America. From Wagner, moreover, Langstroth got his first knowledge of the work of Dzierzon of Silesia, the foremost contemporary European observer of the life of the hive, a man of just about his own age, whose recent book, *Theorie und Praxis des neuen Bienenfreundes* (1848), Wagner had translated into English.

The acquaintance with Wagner came about through the good offices of the Reverend Dr. Joseph Frederic Berg, pastor of the German Reformed Church in Race

Street, Philadelphia. He was a learned man whose general information is said to have been encyclopedic.[1] In the summer of 1851 Dr. Berg read in a Philadelphia newspaper that a neighbor of his, the Reverend Mr. Langstroth, had discovered or contrived a method by which bees could be made to work in large glass "observing hives," even in broad daylight, and prevented from obscuring the glass with propolis. His interest was aroused, for a reason which will presently appear, and he sought and obtained an invitation to visit this neighbor's apiary. There he saw bees working in hives which had side walls consisting of two panes of glass separated by a dead-air space. He saw more than that, for Langstroth took him about and demonstrated his methods of operation. Years afterwards Langstroth, recalling Dr. Berg's visit, told of it as follows:

From him I first learned of the existence of such a person as Dzierzon, and of the great attention he had attracted by his successful management of bees. Before Dr. Berg communicated any particulars of Dzierzon's methods I showed him my hives, and explained my system of management. He found our hives to differ in some very important respects; but he was greatly astonished at the remarkable similarity in our methods of management, as my investigations had evidently been conducted without even the slightest knowledge of Dzierzon's labors. He informed me that Mr. Samuel Wagner, cashier of a bank at York, Pa., had made a translation of Dzierzon's work on bees, the loan of which he procured for me.

[1] See the sketch of Berg's life by Frederick T. Persons in the *Dictionary of American Biography*.

No words can express the absorbing interest with which I devoured this work. I recognized at once its author as the *Great Master* of modern apiculture. His discovery of parthenogenesis threw a flood of light upon the profound mysteries in the physiology of the honeybee, which had so perplexed observers from the time of Aristotle, and which even Swammerdam, Réaumur, and Huber had failed to solve. I soon perceived that I had been anticipated in more than one important discovery, and that he was well acquainted with the fact, with all its practical results, that bees without a queen build only drone-combs. Artificial processes, which I had supposed to be all my own, and which I had practiced on a comparatively small scale, had been conducted by him so largely and so successfully as to secure special recognition by the king of Prussia, at whose request his book was written.[2]

That statement gives a hint of the stirring effect which this adventure must have had. Heretofore he had pursued his investigations out in front alone, with little to whet his courage or to gauge his progress but the records of past exemplars. Now he had found a patron and counselor in Wagner and a fellow worker in Dzierzon. For there is no sign of jealousy in his admission that Dzierzon had anticipated some of his discoveries and processes. Rather he appears to have taken his own measure and thereby gained assurance of the rightness of his aims and the merit of what he alone had done. He found himself in good company.

How came the cashier of a bank in a small Pennsylvania town to be so eager and prompt in translating a scientific work? A short account of Samuel Wagner's life

[2] Reminiscences, *Gleanings in Bee Culture*, xxi, 116–18.

and interests will give the answer.[3] He was born at York
in 1798, the son of a minister of the German Reformed
Church. German was the language of the family and of
its neighbors, and Wagner spoke no other until he was
almost ten years old. He attended a parochial school and
finished his formal education at the York county academy. After several years of mercantile occupation he
became the owner and manager of two newspapers in
succession, the York *Recorder* and the Lancaster *Examiner.* He sold the latter when he was appointed cashier of
the York bank. Meanwhile he had become a fascinated
student of bee culture, just how does not appear, but the
process can be imagined by anybody who has felt a similar fascination. He had pursued the study by way of reading rather than practice and experiment as Langstroth
had. His knowledge may therefore have been "academic," as they say, but Langstroth testified that it was
wide and exact. Knowing two languages, Wagner had
made use of both. There were German books about beekeeping and he had imported them. There were even
German bee journals, two of them, and he had been
reading them both from their beginnings.

Visiting York in September, 1851, a few weeks after
his call on Langstroth, Dr. Berg told Wagner about this
extraordinary beekeeper whom he had talked with in
West Philadelphia and about the contrivances and meth-

[3] The data are found in two obituary articles, the one written by his
son and the other by Langstroth, which appeared in the *American Bee Journal* in March, 1872.

ods which he had seen in use there. They agreed that Wagner must go and see for himself, but it was not until almost a year later, in August, 1852, that he was able to do so. On that visit he missed Langstroth for some reason, probably illness and seclusion, but he went to the apiary, opened and inspected hives, and noted their interior arrangements. What he saw, of course, were the movable frames which Langstroth had by this time substituted for bars.

Then Wagner made a decision, at a cost of no little self-sacrifice, which had most important consequences for Langstroth and for beekeeping generally. He had learned that Dzierzon's improved methods were making bee culture a highly prosperous business in Germany and he had hoped that they might, when made known through his translation, do as much for the industry in America. He had even begun negotiations with a publisher. But now he gave up that project. Instead of going on with it he urged Langstroth to write a book himself, describing his new hive and his system of management. He declared the movable-frame hive to be far better than Dzierzon's, and he professed quite as much respect for Langstroth's methods as for those of the Silesian. Such a book as he now advocated would, he believed, do more for practical beekeeping in America than could anything from abroad. For his part he would put all his store of information at his new friend's service. "Seldom," Langstroth thought, "do we find such an admirable example of rare magnanimity and disinterestedness." [4]

With new courage, therefore, he put the book first in his plans for the immediate future. In November, 1852, he went to Greenfield to live. He left his family in Philadelphia. His wife found a place as teacher in a girls' school and the two daughters were enrolled in that school. He made his home—for the next six years, as it turned out—with his sister Margaretta, who had married Almon Brainard,[5] a substantial citizen of Greenfield, in 1848. He began the writing of his book at once. His wife helped him as amanuensis and he needed her help, for much of his writing was a scrawl which only he or she could decipher. He sent his manuscript, a few pages at a time, to her in Philadelphia and she made a legible copy for the printer. His work was done with remarkable speed. Within a few months he assembled the material and accomplished the writing of a duodecimo book of 384 pages. "With the pecuniary aid of [his] kind brother-in-law" [6] he had the type set and the printing done by C. A. Mirick in Greenfield, and in May, 1853, the firm of Hopkins, Bridgman & Company

[4] Reminiscences, *Gleanings in Bee Culture*, xxi, 250–51.

[5] A native of Randolph, Vermont, and a graduate of Hamilton College in the class of 1825. He studied law, was admitted to the bar, and began practice at Greenfield in 1829. There he held numerous offices of trust. He was at various times a member of the school board, paymaster of militia, and secretary of the Franklin county mutual insurance company, and he was county treasurer and register of deeds from 1842 until 1856, when he was elected to the state senate.

[6] Presumably Almon Brainard, whom he has just mentioned by name in this section of his Reminiscences. Two other brothers-in-law gave him substantial help at times. See Chapter X.

of Northampton, Massachusetts, published the first edition—"a small edition" in the author's account of it—of *Langstroth on the Hive and the Honey-Bee: A Bee Keeper's Manual.*

This was the first soundly scientific and yet simple book on the physiology and habits of the honeybee and the principles of its culture that had appeared in the United States, and it had no rival for many years. It imparted information never before made available in English. Its practical instructions were the fullest and the best yet assembled and offered to beekeepers.

An introduction related the author's experience in the rearing and care of bees and recounted the series of experiments which had led to his invention of the movable-frame hive. The next chapter, "The honeybee capable of being tamed or domesticated to a most surprising degree," told how to manage bees without angering them. Other chapters dealt with the physiology of the queen, the drone, and the worker; the nature of comb, propolis, and pollen; protection against extremes of heat, cold, and dampness; ventilation of the hive; swarming and hiving; enemies of the bee; treatment of swarms that have lost their queens; robbing and its prevention; union of colonies, transferring of bees, and the starting of an apiary; the feeding of bees, and their pasturage. In short, the book told practically all that one need know about the care of bees. Its maxims were backed up with reasons and illustrations drawn from experience or from wide reading. Except for a few lit-

erary excursions, after the fashion of the time, its style
is direct and concise. Its advice for the most part is as
good now as it was in 1853.

A second edition, revised by the author, was brought
out in 1857. In 1858, after living with his sister in
Greenfield for six years, Langstroth gathered his family
together and moved with them to Oxford, in southwest-
ern Ohio. About that time he completed another revision, and in 1859 the J. B. Lippincott Company of Philadelphia published a third edition. The demand for
the book continued and new printings were made from
time to time, but without further revision. Recurrent
attacks of nervous prostration disabled the author for
such a task, much to his regret, for he would have liked
to keep his book abreast of new processes and appliances,
such as artificial foundation for combs, and the centrifugal honey-extractor. In 1872 he was severely injured in
a streetcar accident and was laid up for months. He had
hardly recovered from that injury when he had a slight
stroke of paralysis. The effects of that wore off in a year
or so, but afterwards his nervous malady disabled him
for longer periods of time.

It was not until the 1880's that he arranged to have
his book again revised, this time not by himself but by
others, namely, Charles Dadant (1817–1902) and his
son Camille P. Dadant of Hamilton, Illinois. The elder
Dadant had come to this country from France, had settled at Hamilton, just across the Mississippi River from
the town of Keokuk, Iowa, and there had built up a suc-

[Title-page of First Edition]
LANGSTROTH

ON THE

HIVE AND THE HONEY-BEE,

A

𝕭𝖊𝖊 𝕶𝖊𝖊𝖕𝖊𝖗'𝖘 𝕸𝖆𝖓𝖚𝖆𝖑,

BY

REV. L. L. LANGSTROTH.

EVERY GOOD MOTHER SHOULD BE THE HONORED QUEEN OF A HAPPY FAMILY.

NORTHAMPTON:
HOPKINS, BRIDGMAN & COMPANY.
1853.

cessful apiary. Both he and his son had made themselves known as competent writers on bee culture, the elder especially in France, where a book and numerous articles of his had been published.

On the advice of a friend and fellow beekeeper in Cincinnati, Charles F. Muth, who knew the Dadants well, Langstroth asked them to help him in preparing a new edition of *The Hive and the Honey-Bee*. A correspondence begun with them in 1881 was so often interrupted and so long delayed by his nervous breakdowns that it was concluded only after eight years.[7] Meanwhile the Dadants had been obliged to take upon themselves virtually all of the labor of revision. The financial arrangements had become so complicated by reason of the author's physical incapacity for business that there was only one fair and practicable way out: the Dadants themselves published the new edition. They brought it out in 1889. They have published numerous subsequent editions.

The book has been translated into five continental European languages. Not long after he finished his work on the revision of 1889 Camille Dadant completed a French translation which was published in 1891. It has had such a wide circulation in French-speaking countries that new printings were made in 1896, 1908, and 1923. A Russian translation was published in 1892 and was reprinted in 1909, 1925, and 1929. A Spanish ver-

[7] Copies of Langstroth's letters and copies of some of the Dadants' letters to him in the course of this long correspondence are preserved in the Cornell beekeeping library.

sion came out in 1915 and again in 1924. Then the book was published in Italian, 1928, and in Polish, 1930.[8]

§2. WRITING FOR THE JOURNALS

The first edition of *The Hive and the Honey-Bee* carried to American beekeepers the first public suggestion of their need of a periodical publication devoted to their interests. In his introduction (on p. 23) the author noted that there were in Germany two such journals, one of them published for more than fifteen years, and he made this prediction: "There is now [1853] a prospect that a Bee Journal will before long be established in this country. Such a publication has long been needed. Properly conducted, it will have a most powerful influence in disseminating information, awakening enthusiasm, and guarding the public against the miserable impositions to which it has so long been subjected."

Evidently Langstroth and Samuel Wagner had been talking with each other about ways and means of getting out such a publication. There can be no doubt about it, in the light of subsequent events, for Wagner did establish the first one in this country, the *American Bee Journal,* in 1861, and from the start he had Langstroth for a contributor.

Before that came about, however, Langstroth made

[8] Maurice Dadant, present Editor of the *American Bee Journal,* kindly furnished this list of foreign-language versions.

his entry into journalism. In 1854, the year after his book appeared, he had two articles in the *American Agriculturist*, then in its twelfth year. Wagner launched the *Journal* at a time which proved unfortunate, the first year of the Civil War. After that year it was suspended and did not resume publication until 1866. The first volume contained several articles written by Langstroth. During its suspension, in the course of five years, he contributed nine articles to the *Country Gentleman*. After the *Journal* was revived he wrote for it as regularly as his health permitted.

Samuel Wagner died in 1872, and after that event—deeply mourned by Langstroth—the *Journal* had a succession of different owners and editors until 1912, when it was acquired by the Dadants, who still edit and manage it. In 1873, at a time when the *Journal* seemed to be declining from its former high standard, Amos Ives Root[9] of Medina, Ohio, established *Gleanings in Bee Culture*. Both the periodicals wished to publish Langstroth's occasional articles, which were now eagerly read by beekeepers, and the two proprietors came to an unusual agreement with each other. In 1874 they arranged with him for the publication of his contributions in both journals simultaneously. That was done for years.

Readers came to expect from him now and then an essay in which a sprightly or jocular style lent interest even to a faithful recital of biological fact. In one such

[9] The subject of an interesting sketch by John I. Falconer in the *Dictionary of American Biography*.

article, published in March, 1888, in the *Journal* and in *Gleanings,* he presents the "poor slandered Drone," who pleads for and obtains "his day in court." That article is reprinted here, in full except for a few subordinate headlines. The Drone speaks:

Virgil, who was a great poet, but not enough of a practical beekeeper to know a laying from a virgin queen, was the first writer of much note to have his fling at me. To him I was only an idle knave, born to consume the fruits of others' labors, and deserving no better fate than death, by ignominious expulsion from the industrious commonwealth. Ever since he so grossly libeled me, to compare one to a drone is the most orthodox form of denunciation for laziness, gluttony, and what has been called "general cussedness."

Now, I am proud to say to this court that I can disprove every charge brought against me, by simply proving that, to the best of my ability, I fulfill the express purpose for which I was born. Surely no creature can do any better than this, and excuse me for thinking that few men do as well.

If any of my enemies had authority to call the roll of my demerits, he would surely begin by accusing me of being too *lazy* to gather any honey. But an expert in points of this kind could remind him, that, if he examines my proboscis, he will see that it is much too short for sipping nectar from the opening flowers. I am free to admit that I make no wax; but even Cheshire himself, whose microscopes have fairly turned me inside out, will tell you that I have not a single wax-secreting gland, and am also without those plastic, trowel-like jaws which enable the worker-bee to mold the wax into such delicate combs.

Now, do not insinuate that I might at least employ some of my leisure time in gathering pollen! Can you not see that my thighs have no basket-like grooves in which it could be packed, and are quite destitute of the bristles by which the workers hold the pollen in place?

No doubt you have often denounced me as a big, hulking coward that leaves to the women the whole defense of the state. Are you not aware that I have nothing to fit me for acting on the offensive? Would that I had one proportioned to my bulk! If only that I might make proof of it upon all who berate me for not accomplishing impossibilities! I am not at all ashamed to admit that I spend the most of my time, not given to eating, either in sleeping or what you are pleased to call listless moping about the hive. Has it never occurred to you that, if I should try to assume the restless activity of the worker-bee, I could be nothing better than a meddlesome busybody, perpetually interfering with the necessary business routine? I guess the silly meddler who put me up to such nonsense ought more than once to have had a dishcloth pinned to him, to teach him not to bother the women in their work!

I am sorry to number Shakespeare among those who have misconceived me, by calling me "the lazy, yawning drone"; but, as one of my maligners has likened me to Falstaff, I may be allowed to quote, in my own defense, what this great braggart, when accused of cowardice, says of himself to the prince: "Was it for me to kill the heir-apparent? should I turn upon the prince? why, thou knowest I am as valiant as Hercules: but beware instinct; the lion will not touch the true prince. Instinct is a great matter; I was now a coward on instinct. I shall think the better of myself and thee during my life; I for a valiant lion, and thou for a true prince." I lie not, like the false knight, when I say that what you call my laziness is a matter of pure instinct.

With all your boasted reason, you seem to have overlooked the doctrine of conservation of forces. You upbraid me with consuming so much of the precious honey, to the gathering of which I contribute nothing! Well! if I made a single uncalled-for motion, would not that necessitate an extra consumption of food? What better can I do, then, than to keep as quiet as possible? There is nothing either outside or inside of the hive which calls for any other line of conduct, until the young queens are on the wing; and as they do not sally forth until long after noon, why should I go abroad any earlier? I can assure you that if bridal excursions were in order as many hours in the day as the flowers secrete honey, no worker would ever be earlier to rise, or later to go to bed than myself.

I an idle, lazy, listless lounger, forsooth! Does any one wish to witness the most perfect embodiment of indefatigable activity? Let him then look at me, when, at the proper time, with an eager, impetuous rush, and a manly, resonant voice, I sally from the hive! See with what amazing speed I urge what our old friend Samuel Wagner called my "circumvoluting" flights! For aught you know, I may cover greater distances in describing these vast circles than the busiest worker in the longest summer day. There is great need, then, that I should be abundantly provisioned for such exhausting excursions; and it is only a law of nature that, on my return from them, all that I carried out with me should be found to have been used up. If you taunt me either for the full or the empty stomach, I merely ask you if you have never heard of honeymoon trips among your own people, which began with extra-full purses, to end only with uncomfortably light ones.

To cap the climax of your abuse, what savage delight you take in seeing the worker drive me from my pleasant home! and how glibly you can moralize over what you call a righteous judgment upon a life spent in gluttony and inglorious ease! Just as if you did not know that the whole economy of the bee-hive is founded upon the strictest principles of utilitarianism! Is not a worker-bee, when disabled by any accident, remorselessly dragged out to die, because it can no longer contribute to the general good? Even so exalted a personage as the queen mother herself, as soon as it is plain that her fertility is too much impaired, has a writ of *supersedeas* served upon her, in favor of one of her own daughters.

Knowing well the law under which I was born, I urge nothing against being put to death when Shakespeare's "pale executioners" deem the day of my prospective usefulness to be over. Truly, the sword of Damocles is suspended over my head; and from the hour of my birth till that of my death it may fall at any moment. Many bitters are thus mingled with my sweets.

I have time to mention only one more. While I know that most of the young queens come safely back from their wedding excursions, I can not help foreboding the worst, when I see that no drone ever returns to tell of his experience.

I will close my defense by reminding you how the good father of the great Scotch beekeeper, Bonner, showed his appreciation of our persecuted race. It was his custom to watch every year for the first flying drone. Its cheerful hum so filled him with delight, as the happy harbinger of approaching swarms, with their generous harvests of luscious sweets, that he called an instant halt on the work of his busy household, and devoted the rest of the day to holiday feasting. The patron of the drones ought for ever to bear the honored name of "Saint Bonner."

THE DECISION OF THE COURT. This court having heard the defense of Sir Drone, pronounces him to be innocent of each and every one of the misdeeds alleged against him. It only regrets that it can not inflict adequate punishment upon his slanderers. Alas, my poor fellow! the lies against which you protest have had so many centuries the start of your true story that you may well despair of ever overtaking them in your short lifetime.

MORALS. From the plea of the drone, many good morals might be drawn; for, "As he is guilty, that shooteth arrows and lances unto death, so is the man that hateth his friend deceitfully, and, when he is taken, saith, 'I did it in jest.'"

VII

A New Era in Beekeeping

THE MOVABLE FRAME was the very key needed to open the door to successful beekeeping on a large scale. The inventor himself was confident of that. He had expressed that faith in the careful entry which he made in his Journal on the day when the idea of the precisely spaced frame had come to him, October 30, 1851. The practical value of his invention to a great industry was almost beyond estimate.

He filed application for a patent on January 6, 1852; it was granted on October 5 of the same year.[1] In November he went to Greenfield, Massachusetts, where he had lived from 1839 until 1848, there to make his home with his sister [2] and to solicit the help of friends with means and influence in introducing the new hive. He eventually got Dr. Joseph Beals, a dentist of Greenfield and former parishioner of his, to put up capital for man-

---

[1] It was Patent No. 9300, granted for a period of fourteen years. A reissue, No. 1484, was granted on May 26, 1863. The reissue was asked for to enable the patentee to amplify and clarify his original application so that, in the event of litigation, which was then impending, all the pertinent information would be on record. An extension of seven years was granted on October 4, 1866. The original patent, on vellum, is filed in the Patent Office, for its surrender was a condition precedent to the reissue.

[2] See above, Chapter VI, Note 5.

ufacture and promotion, and gave him in return a half-interest in the patent. The two proprietors then sold to Roswell C. Otis of Kenosha, Wisconsin, the patent right for "the Western States and Territories."

The first edition of Langstroth's book, *The Hive and the Honey-Bee*, which came out in March, 1853, just when he and his associates were beginning to promote the introduction and sale of the new hive, gives an idea of how they were going about it. The book contained a conspicuous advertisement of L. L. LANGSTROTH's MOVABLE COMB HIVE. "Each comb in this hive," it began, "is attached to a separate, movable frame, and in less than five minutes they may all be taken out, without cutting or injuring them, or at all enraging the bees." Various necessary operations which the new hive made simpler and more certain were then enumerated, and the statement went on as follows: "That the combs can always be removed from this hive with ease and safety, and that the new system, by giving the perfect control over all the combs, effects a complete revolution in practical beekeeping, the subscriber prefers to *prove* rather than assert. Practical Apiarians and all who wish to purchase [patent] rights and hives, are invited to visit his Apiary, where combs, honey and bees will be taken from the hives; colonies which may be brought to him for that purpose, transferred from any old hive; queens, and the whole process of rearing them constantly exhibited; new colonies formed, and all processes connected with the practical management of an Apiary fully illustrated and explained." With a statement of prices and terms, the advertisement concluded:

The hive and right will be furnished on the following terms. For an individual or farm right, five dollars. This will entitle the purchaser to use and construct for his own use on his own premises, as many hives as he chooses. The hives are manufactured by machinery, and can probably be delivered, freight included, at any Railroad Station in New England, or New York, cheaper than they could be made in small quantities on the spot. On receipt of a hive, the purchaser can decide for himself, whether he prefers to make them, or to order them of the Patentee. For one dollar, postage paid, the book [3] will be sent free by mail. On receipt of ten dollars, a beautiful hive showing all the combs, (with glass on four sides,) will be sent with right, freight paid to any railroad station in New England or New York: a right and hive which will accommodate *two* colonies, with glass on each side, for twelve dollars; for seven dollars, a right and a well made hive that any one can construct who can handle the simplest tools. In all cases where the hives are sent out of New England or New York, as the freight will not be prepaid, a dollar will be deducted from the above prices. Address L. L. LANGSTROTH, *Greenfield, Mass.*

In the introduction to his book the author and inventor made a statement of further claims for his hive and his methods, beginning it by depicting, probably without exaggeration, the sad state of bee culture in America at that time:

The present condition of practical beekeeping in this country, is known to be deplorably low. From the great mass of agriculturists, and others favorably situated for obtaining honey, it receives not the slightest attention. Notwithstanding the large number of patent hives which have been introduced, the

[3] Presumably *The Hive and the Honey-Bee*. The advertisement does not make this any clearer.

ravages of the bee-moth [4] have increased, and success is becoming more and more precarious. Multitudes have abandoned the pursuit in disgust, while many of the more experienced, are fast settling down into the conviction that all the so-called "Improved Hives" are delusions, and that they must return to the simple box or hollow log, and *"take up"* their bees with sulphur, in the old-fashioned way. In the present state of public opinion, it requires no little courage to venture upon the introduction of a new hive and system of management; but I feel confident that a *new era* in beekeeping has arrived, and invite the attention of all interested, to the reasons for this belief. A perusal of this Manual will, I trust, convince them that there is a better way than any with which they have yet been acquainted.

If that last paragraph seems to paint too dark a background for the "new era" which Langstroth was ushering in, such an impression may be removed by reading a letter in which Jared P. Kirtland, an eminent naturalist,[5] told his own experience of the ravages of the bee-moth, or wax-moth, in different parts of this country, and expressed his conviction that an effectual control of that pest had been made possible only by the introduction of the Langstroth hive. His letter was written only a few years afterwards, under date of Cleveland, Ohio, February 19, 1859, and was addressed to Langstroth himself. It was as follows:

[4] *Galleria mellonella,* now commonly called wax-moth.

[5] Born at Wallingford, Connecticut, in 1793, and educated at Yale. His father moved in 1803 to the Western Reserve and the son lived in that part of Ohio after 1825. In 1843 he was one of the founders of the Cleveland Medical College, where he was professor of the theory and practice of medicine until 1864 and emeritus professor until his death in 1877. What he says in this letter of his interest in the wax-moth as early as 1806, when he was only about thirteen years old, is surprising, but he is said to have discovered parthenogenesis in the moth of the silkworm when he was only fifteen. See Frederick C. Waite's sketch of his life in the *Dictionary of American Biography.*

DEAR SIR: Until 1805 the honey-bee flourished in the United States. At the commencement of the present century, a majority of the farmers and mechanics in the State of Connecticut cultivated the bee. Few, if any, unfavorable contingencies interfered with that pursuit; the simplest form of box hive was employed; though occasionally, a hollow gum, and in a few instances, the conical straw skep supplied their place.

In autumn, the weak colonies, and such of the old as were depreciating in value were destroyed by fire and brimstone. The honey thus obtained was sufficiently abundant to supply the demand; hence, in those days, caps, drawers, and side boxes for robbing bees were not employed.

In the spring of the year 1806, I read an article in the Boston *Patriot* describing the miller and worm [moth and larva], and their depredations, and representing them as of recent appearance in the vicinity of that city. A few months subsequently, a neighbor informed me that they were depredating extensively on his colonies; and within two years of that time, four-fifths of all the apiaries in that vicinity were abandoned.

Since that period, a succession of patent hives, whose originators were ignorant of the moth, have appeared as its auxiliaries, and the two combined have nearly exterminated the bee from that section of the country. The efforts of a few individuals, of more than usual perseverance and ingenuity, were occasionally attended with limited success.

In the summer of 1810, I resided in the county of Trumbull, Ohio. The moth had not reached this part of the country, and bee culture was extensively pursued, and with a success I have never witnessed elsewhere. The rich German farmers were on a strife to excel each other in the number of their colonies. Two or three hundred they frequently attained.

In 1818 I again visited that county, and permanently located there in 1825, and at both periods found that pursuit still prospering. In August, 1828, while visiting a sick family in Mercer County, Pennsylvania [adjoining Trumbull County, Ohio], I observed that a large apiary was suffering severely from the attacks of a worm. The proprietor informed me that it had made its appearance for the first time the present season. Within another year, it spread all over northern Ohio, and in the winter of 1831–32, I learned from members of the Legislature that it had reached every part of our State. Similar results followed its progress as in the New England States.

Until the introduction of your system of movable frames, no successful means of counteracting its ravages were devised. I am happy to say that by the use of your hives, I have not had the least difficulty in meeting it. With great respect, etc.,

L. L. Langstroth.   JARED P. KIRTLAND.

It is now well known that the wax-moth does not ravage strong and healthy colonies of bees, and Langstroth himself, in the very first edition of his book (p. 263), emphatically said just that. What made the movable-frame hive a safeguard against depredation by the moth was the perfect control that it gave the careful keeper over conditions within the hive, enabling him to detect promptly an indication of anything wrong there—a lack of stores, or a brood disease, or the loss of the queen—which might, if not remedied, so weaken and discourage the colony as to make it an easy prey.[6]

Among those who bought the new hive with the patentee's license and put it to use in large apiaries were some whose patronage was especially valuable because they were looked up to in their respective parts of the country as men of judgment. These included W. W. Cary of western Massachusetts, Richard Colvin of Baltimore, Charles Dadant of western Illinois, Adam Grimm of Wisconsin, and Moses Quinby of the Mohawk Valley.

6 Phillips, *Beekeeping*, 1928, pp. 436–7.

Their adoption of the hive and their testimony to its merit gave an impetus to its sale.

By no means all of those who acquired patent rights held to the shape of the model hive which the patentee sold. That was an oddity to most experienced beekeepers, being unusually low and broad. They had hives of different patterns which they were used to working with, and while they were careful to keep the essential element of the Langstroth system—the hanging frame bounded by the bee space—they fitted such frames to hives of their own preference. No doubt buyers had a right to do so. But their innocent alterations of "the Langstroth hive" became one of the causes of misunderstanding and confusion of thought which for many years afterwards clouded the question of Langstroth's legal and moral right to the product of his invention.

Unfortunately, the inventor himself had sown seeds of confusion. The hive which he offered to practical beekeepers retained some peculiarities of design which had served his purpose of observation but did not now make sense to the simple producer of honey. There was something queer about its dimensions. It was 10 inches high, 14⅛ inches wide, and 18⅛ inches long. Why that odd eighth of an inch? The question takes us back to West Philadelphia in 1851.

In his experiments with bar hives Langstroth had come to use a low and broad box for reasons which he firmly believed to be scientifically and practically sound. In 1851, just before he learned to substitute frames for the bars, he had settled upon the use of a hive 6 inches high and 18⅛ inches square. And this is why. He made his hives with double walls of glass. A standard sheet of glass, 12 x 18 inches, gave him two panes 6 x 18. To provide a dead-air space the panes of each wall were separated by inserting at the ends narrow strips of veneer one-sixteenth of an inch thick. Hence the odd one-eighth. When he made his first movable-frame hives, in 1852, he reduced the width by 4 inches with a view to better wintering. The next year, at Greenfield, he increased the height from 6 to 10 inches and made the walls of wood instead of glass. But the odd eighth of an inch was retained.

Argument went on for years about the proper form and size of the hive. Not so much was heard of it after the factories set their standards. And those standards, though they may vary in some particulars, are all alike in one thing: they are based upon the use of spaced and interchangeable frames.

The truth that has prevailed over all argument is that the principle of Langstroth's frame, if not the particular form of hive that he designed, came into such wide use as to make other types of hive obsolete. He did indeed begin a "new era" in beekeeping. Virtually all of the processes of the industry were changed for the better and were standardized wherever the Langstroth frame came to be used. That use is now so general in North America as to underlie the whole business of producing honey for the market. The pound section, a common article of sale since 1875, is a refinement of the frame. All machinery now made for the extraction of honey in commercial quantities is adapted only to the handling of framed combs, because they can be put through the extracting process in the cleanest and most economical way.

Hives made on the Langstroth principle, that of the hung frame surrounded by the bee space, are used all round the world, though known by various names other than the inventor's.[7] A judgment of the world-wide value of the invention was given by an able European, Philippe J. Baldensperger of Nice, who may fairly be called the dean of French apiarists,[8] in a letter written in 1925, responding to an inquiry from the author, and including a graceful tribute of respect to Langstroth's memory. Mr. Baldensperger wrote as follows:

My first knowledge about the Rev. L. L. Langstroth dates back to the beginning of the eighties of last century, when Frank Benton [an American], in 1881, talked to me about the great beekeeper. When we raised queens in Syria, Benton talked about all the most notable bee-men—of Huber and his leaf-hive especially, which was in our minds the prototype of the bar-frame hive. We discussed the invention of Langstroth and his complete frame isolated all around from the hive-body; and the top-bar of Dzierzon's hive, which was just short of being

[7] For particulars see Phillips, *Beekeeping*, 1928, chap. ii, "Apparatus," especially pp. 25–8 and 33–5.

[8] Mr. Baldensperger was for years editor of a journal of beekeeping published by the Société des Alpes-Maritimes, of which he was for some time president. He gave further evidence of his regard for Langstroth's memory in 1929, when he persuaded a number of societies of beekeepers in France to raise a fund of money for the purchase and gift to the Cornell beekeeping library of a fine collection of books which had belonged to the late Robert Hommell, author of one of the best French books on the culture of the bee.

a fixed hive; of the Berlepsch hive, which resembled Langstroth's invention, but was to be in conformity with the German idea —breech-loader, or a hive opening at the back, to facilitate the cutting of the comb sticking to the walls of the hive. Certainly Benton was in favor of the Langstroth hive, and being his disciple, I was a thorough Langstrothian. In France, the Debeauvoys hive in the earlier part of the nineteenth century was in favor, as being a great advance on the skep, but the hive being too complicated was abandoned, and the skep remained in favor. Charles Dadant, a Frenchman living in Illinois, was the first to call attention to the Langstroth invention amongst popular beekeepers in France, and he had to fight Hamet, founder and editor of *L'Apiculteur*, in the sixties of last century. Dadant's hive being a modification of the Langstroth hive, the name of the Frenchman was and remains more widely known. By the Root-Langstroth hives and the Root goods, Langstroth's name was refreshed to the memory of Frenchmen, although again, Root's name, as Dadant's, became more popular.

The bee-journals of 1895 and '96 talked about the great bee-master's life and death. Moving continually from the East to the West, I have not kept any of the French bee-papers of those days.

During the period of my Palestine beekeeping, 1880–1892, we called our hives Langstroth system, though they only resembled the master's hives inasmuch as the frames were completely movable. We had to fight the Palestine pear-shaped tube and the German "breech-loader" and put forward our pastoral hives, which were, on account of the small frames—10 by 11 inches—very handy and easy to be transported on camel-back, the then only accessible way. We called them Langstroth hives by deference to the first inventor before we ventured to call them "Baldensperger pastoral hives" to distinguish them from the many names springing up.

In our days of practical money-making, rather than anything else that has taken hold of humanity, the names of those persons who have mostly contributed to swell our purses will be more remembered than the real originator of an idea. Nevertheless, the name and memory of the Rev. L. L. Langstroth will ever be cherished the world over.

In the old countries—Europe and Asia—every man sticks to his own native laws and language, and though many acknowledge some superiority to a foreigner, the native name is best known and venerated. Dzierzon, who discovered the law of parthenogenesis, has the greatest name in the German beekeeper's memory; so has blind Huber a great name in French society.

Langstroth and Huber were always foremost in my imagination, and when I had a few dozen hives only, to commemorate the great beekeepers, each hive had a name, and the two above-named were always at the top.

French books on beekeeping sometimes just mention the name of Langstroth without giving many items on his invention. But the Langstroth hive revolutionized the beekeeping world, whether the name of Langstroth is ever heard or not.

VIII

The Golden Bee Imported

BESIDES improving the hive and enriching the literature of bee culture, Langstroth rendered a third great service to beekeeping in America by sharing in long-continued and finally successful efforts to import Italian bees to this country and naturalize them here. "Their introduction marks an important milestone in American apiculture, almost equal to the invention of the movable-frame hive." [1]

Different races of the honeybee have been recognized since ancient times. Aristotle describes three. Vergil distinguishes two, the members of the one race dirty-colored, "the others splendid, their glowing bodies adorned at regular intervals with spots of gold." These latter he declares to be the better stock. [2] Columella, in the next century after Vergil, describing this preferred bee, observes that its disposition is more peaceable than

---

[1] Phillips, *Beekeeping,* 1928, p. 211, in a chapter entitled "Races of Bees."
[2] *Georgics* IV. 98–100:

> Elucent aliae et fulgore coruscant
> ardentes auro et paribus lita corpora guttis.
> Haec potior suboles.

that of others. Those ancient descriptions permit of no doubt that the modern Italian bee is of the same superior race. In some favored parts of Italy the strain must have kept its golden purity throughout the ages.

In the nineteenth century keepers of bees in other European countries learned about this bee of Italy and tried to import it. One of the earliest records of such an attempt was published in 1848 in the Eichstätt (Bavaria) *Bienenzeitung,* relating the experience of a Swiss officer, a beekeeper by avocation, one Captain Baldenstein. In the course of the Napoleonic wars he was stationed in the Valtellina, near Lake Como. There he noticed honeybees that were brighter in color than those which he had known, and he observed that they were more industrious. Years afterwards, in 1843, retired and at leisure in Switzerland, and taking up his hobby again, he remembered those bees that he had seen in the Valtellina and resolved to get some of them if he could. He sent two men, who bought a colony for him and brought it over the mountains to his home. He tried to propagate the strain, but he was not able to keep it pure. He did learn enough about it to prove its superiority.

Jan Dzierzon of Silesia, reading of Captain Baldenstein's evidence, determined to make another effort himself. He first made sure that the Italian bee's cells were of the same size as those of the common kind. Then, with the help of the Vienna office of the Austrian agricultural society, he succeeded, in February, 1853, in transporting a colony from Mirano, near Venice. He was almost immediately successful in rearing numbers of Italian queens.

It was not long before enterprising American bee-keepers, aware of what had lately been done in Europe, undertook to replace the common black bee of this country with the golden stock, by all accounts so much superior. There were sporadic attempts to import colonies, but they were not at first successful. As early as 1855 the energetic Samuel Wagner arranged for a shipment. It was made, but the bees perished on the voyage. In the spring of 1859 Langstroth and Wagner, together with Richard Colvin of Baltimore, after some preliminary correspondence ordered a colony from Dzierzon, only to learn after long waiting that their order had never been delivered.

About this time the United States Division of Agriculture (then administered by the Commissioner of Patents) undertook the importation of Italian bees. The Commissioner employed S. B. Parsons of Flushing, Long Island, an agent of the Division, to procure a number of colonies in Europe.[3] Parsons wrote to Langstroth, who was then living at Oxford, Ohio, asked him for advice about breeding and disseminating the Italian bees, and invited him to go to Flushing. Langstroth went, and was there in time for the arrival of the first shipment. He remained for almost two months, endeavoring to save as many as possible of the precious queens.

[3] For an account of these and other early importations see Phillips, *Beekeeping*, 1928, pp. 211–12.

He kept in his Journal a record of his patient efforts. An extract, covering a part of the month of April, 1860, is here published for the first time:

Flushing, April 2, 1860. Examined five stocks of Italian bees imported by Mr. S. B. Parsons, brought in the original gums— hollow sections of trees. Only one contained any living bees. Died from want of skill after they were landed. Put in green-house—many died there. The one with living bees had less than a pint—some sealed brood.

3rd [April]. Put the hive with bees in a dry cellar until I could transfer it.

4th. Cut out the combs and found a living queen. I never handled anything in my life with such care. If the queen had been killed, I should have felt worse than any regicide ever felt, for they *mean* to kill royalty. Strange that so small a creature should be capable of producing so exciting an effect! I carefully put her in a queen case, cut out the only comb that had any brood, and put the small colony in the cellar.

5th. I rose very early, fearful that the queen might have been chilled, and found that the bees had left her. I took her out of the cage with fear and trembling. She was stiff and could hardly move. I warmed her with my breath and returned her to her colony, which I kept in the cellar.

7th. Transferred a colony of common bees and removed queen—gave them Italian in a cage. Near sunset, let out Italian queen very carefully—bees welcomed her. Now I have done all that I could for her and must patiently await the result.

8th. Put hive with Italian for safety on roof of piazza, not knowing that it had been painted within a week. Weather was stormy and a multitude of bees died within the hive. Removed to another location, bees still dying.

10th. One half or more have died.

11th. Bees doing well now. Saw queen all right. Left in P. M. for Boston.

Returned on the 18th and spent day in New York, looking

for arrival of steamer Arago [?] in the further importation of Italian bees. Left New York for Flushing at 5 P. M. Mr. Parsons had received a message stating that the Arago would be at the dock at 9 P. M., so we took a private conveyance to Hunter's Point, and thence to New York, arriving there at 10 P. M.

The German-Italian with the bees speaks poorly of them. Had 27 in small cases, all put up wrong. Too many saw-clefts to let in light and keep bees in a constant worry. Two said to have the best from 140 queens were kept by the beekeeper in his constant charge. Weather bad, hard winds, bees sick. Last two days many died. The two spoken of, one lively, the other "sick" as the Italian said. Tried to get them off that night to keep out of damp, bad air, but could not get permit for them to leave. Spent the night with Mr. Parsons at the Astor House. Went to bed at 1 A. M., rose at ¼ before 5. Bees in my head, all right! Went to steamer—got my two stocks and started for Grand Street ferry. Came within a few minutes of missing Flushing cars at Hunter's Point. One stock gave no responsive buzz. I opened it before breakfast, and all seemed dead. *Starved*, not "sick." I laid them out on paper and a few came to and greedily took honey. I found the queen and put her in a cage with some other bees, and they soon became very lively. I opened the other box, and the bees flew at once. These two Italian queens were given to common bees, as before, and soon the first one began to lay. For the second, a German queen cage was used. The slide was gone, and she got out and flew. I caught her gently with a sweep of my hand—a very narrow escape.

On the 19th Mr. Bodmer [?], the Italian [?] beekeeper, arrived with the balance of the Italian bees. The bees had been put up badly for the voyage, but if the passage had been two or three days shorter most of them might have arrived safe. In some boxes the combs were loose and had moved about so as to crush bees. One queen was insensible but was revived and seemed quite fair, but died in the night. Of thirty colonies, only seven living queens!

I divided a common stock and made two of it, and gave one

an Italian queen in a queen cage. Mr. Bodmer gave an Italian queen to a very small number of bees in a small box with honey comb and took it to a new place. He does not approve of my mode of giving Italian queens so soon.

On the morning of the 23rd, the small colony with the Italian queen given by Mr. Bodmer left the hive and the queen was lost. The same day he made two more such colonies and gave them many bees,—all right. This morning we looked at the one I made and he feared was wrong, and found this all right. "Melissa good, very good. I happy, very happy." I never saw a human being more delighted. Says it is the best queen out of 140—the one I saved from death by starvation! He says that at first he "no like my hive; but now, every day, more, more; he no use Italian-German (Dzierzon) hive any more."

Those weeks of work at Flushing were not wasted. The few queens that were saved enabled Parsons to breed Italian colonies and afterwards to supply many another beekeeper with the nucleus of a similar colony. On Langstroth's recommendation he employed a man from Colrain, near Greenfield, Massachusetts, W. W. Cary, to come to Flushing and take charge of the rearing of queens. Cary was immediately successful, supplying enough to fill numerous orders during that first season. He was skillful also in managing shipments. One order for more than a hundred queens came from the Pacific Coast. They were so carefully prepared for the long journey by Cary and a fellow worker, A. G. Bigelow, that Bigelow conveyed them from New York to San Francisco with a loss of only two. Cary was the first person to send a queen across the ocean in a single-comb nucleus with a few workers in attendance. After seeing

Parsons established in his enterprise Cary returned to Colrain, enlarged his own apiary, and became one of the foremost of reliable breeders of Italian queens.

Langstroth obtained some queens from Parsons and began experiments of his own in the culture of the new strain. Not long afterwards, with the idea of improving it by careful breeding, he opened a correspondence with Dzierzon in Germany and began to import Italian queens from him. A letter of Langstroth to Dzierzon, written under date of Oxford, Ohio, August 22, 1865,[4] is interesting as a specimen of their correspondence and because it shows how industriously Langstroth was working to perfect the breeding of the Italian bee. The text of the letter follows:

I received the bees from Italy, put up by you, and am now having queens from them. I do not expect to get any better queens than those I have imported from your apiary, and those which were imported from Italy before I received yours; still, I am anxious to have queens direct from Italy, to compare with those from which I have previously bred.

I find that the long confinement necessary to bring these queens to America seems greatly to impair their vigor. Of the two queens sent me from your apiary in 1863, one never laid any eggs from which workers could be had; the other proved last spring to be a drone layer, having been only moderately fertile, during the previous season. Of the two sent in 1864,

[4] A copy of this letter, received through the good offices of C. P. Dadant and the courtesy of Leopold Pawkowski, president of an association of beekeepers at the Polish town of Rudnik on Sarr, was read at a meeting held in memory of A. I. Root and L. L. Langstroth at Medina, Ohio, in September, 1926. The original had been given, after Dzierzon's death in 1906, to the museum of the national beekeepers' association of Poland.

one died early this season; the other has proved a prolific queen. I would suggest in preparing bees for a trans-Atlantic voyage that fewer workers be put with the queen, selecting such as are quite young, as they will endure confinement better than the older bees. In one of the boxes sent me last, only two workers were living when the box was opened, and the other two boxes contained each less than a dozen living bees. As I am nearly a thousand miles from New York, it is necessary to have the boxes opened on their arrival, by an expert, and the queens prepared for their long journey. In sending fewer bees, however, do not diminish the supply of honey, as bees in confinement eat enormously. I send them over 1500 miles, and find that unless the weather is quite cool, they bear the journey better when they have but few bees with them. I have successfully sent them by mail, in a small box, with less than a dozen bees, over 1200 miles.

You are well aware that the production of dark or poorly colored queens leads many to doubt the purity of the race. I have discovered the interesting fact that the most beautiful queens can be so treated as to lose all their brilliancy. Take a beautiful queen, just hatched, or very young, wrap her up in some cotton or wool, feeding her three or four times a day, and at the end of a few days she will often be found to have lost all her beauty and to have become almost black. The same treatment will produce no perceptible change in the color of a young worker. This experiment proves that dark color is no proof of impurity in the race.

Although not in the best honey district, I have in several instances obtained nearly 150 pounds of surplus honey from a single stock of Ligurian bees. Some of my friends have obtained even more. I send you my circular containing the points of difference which thus far I have noticed between the Ligurian and the black bees. The most careful measurements show, on the average, no greater length of proboscis in the Italian than in the common variety.

My dear sir, I should esteem it a very great privilege to make your acquaintance, and hope that I may some day be able to visit your apiary.

Since the publication of the last edition of my work on bees, I have ascertained that some of the facts which Professor Siebold, in his work on parthenogenesis, supposed to have been first discovered by himself, were previously known to the celebrated English surgeon, John Hunter. In the Philosophical Transactions of the Royal Society of London, Vol. 82, p. 128 (this article was read to the Society February 23, 1792), Hunter says that his experiments were made in the summer of 1767; he has demonstrated the true use of the spermatheca in the queen bee and other insects, as reservoir for the spermatic fluid. No doubt the learned professor has never met with Hunter's curious observations. We know little more than Hunter on this subject, save that bees pair in the open air; that queens, when once impregnated, remain so for life, and that an unimpregnated egg produces a drone.

To you, sir, the world is indebted for this last discovery, which has thrown a flood of light upon points which have defied the profound acumen of Aristotle and minute researches of Réaumur, Swammerdam, Huber, and later writers.

An entry in Langstroth's Journal of the next month, September, 1865, shows him approving his own persistence in importing Italian queens from different sources: "Important query: Both Dzierzon and myself think we have improved the Italian bee by careful breeding. Have we not lost as much or more by close breeding than we have gained by selection? I presume that we have. For this reason I value so highly the new importations."

Langstroth summed up the results of his comparative study of the black bee and the Italian in a series of articles contributed to the *American Bee Journal* and *Gleanings in Bee Culture* in 1881. Not all of his findings are applicable to beekeeping today because subsequent improvements of equipment and changes of method in the production and marketing of honey have made some of them obsolete. They are pertinent here, however, as evidence of the thoroughness of his observations. And testimony to their practical value at that time was offered in an editorial note by A. I. Root, who wrote in *Gleanings:* "I am quite happy to add that my experience corroborates almost, if not quite, every point that friend Langstroth has made; and inasmuch as this paper is the most exhaustive article we have ever had on the comparative differences of the two races of bees, giving minutely the queer points and peculiarities of each, I feel like, for one, tendering our old teacher a vote of thanks." Langstroth's summary was as follows:

(1) Where late forage is scarce, the Italians stop breeding much earlier than the blacks. (2) The Italians, unless stimulated by judicious feeding, do not resume breeding as early as the blacks. (3) The Italians are much more inclined to build drone comb than the blacks. (4) The blacks are more ready than Italians to work in surplus honey receptacles not closely connected with the main hive. (5) The comb-honey made by the blacks from any light-colored supplies is usually more attractive than that stored from the same sources by Italians. This is owing to the former leaving a larger air-space than the latter between the cappings and the sealed honey. (6) With a queen of the current year, the blacks will hardly ever swarm, while long after the swarming season young Italian queens will often lead off swarms. (7) Black bees are much more sensibly affected by the

loss of their queen than are the Italians. (8) In building, an Italian swarm seldom begins as many combs as the blacks, and therefore works them more completely, squaring them out, as it were, as they proceed. (9) Black bees will readily build, between guide-frames, worker combs, while it is very difficult to get any satisfactory result in this line from Italians. (10) The Italians, both young and old, adhere with much tenacity to their combs when they are lifted from the hive, while the blacks, more especially those newly hatched, tumble off so readily as to annoy the operator by crawling up his clothes, or exposing themselves to be trodden upon.[5] (11) When the hive is opened, the Italian queen and workers are disposed to remain quiet, and when the frames are lifted out, the workers spread themselves over the combs. For this reason, as also from their bright color, Italian queens are readily found, while the blacks, both queens and workers, often fairly race off the combs, and that greatly increases the difficulty of many important operations. (12) Under adverse circumstances, the blacks are far more easily discouraged than the Italians. I soon learned this to my cost when I was obliged to use black bees in making nuclei for rearing Italian queens. (13) The Italians will, in some seasons, from the second crop of clover, build new combs and store them with honey, when black stocks in the same apiary are losing weight. (14) Italians suffer little, compared with the blacks, from the ravages of the bee-moth. (15) Italians are far less likely than blacks to rob or be robbed. Those who have kept only Italians can form but a faint idea of the incessant vigilance required, during the whole working season, to prevent robbing among black bees. (16) The Italians, by their superior energy and greater length of proboscis, will, on an average of seasons, gather much larger stores of honey than

[5] In those characteristics which are noted in the 10th and 11th of these observations, the American black bee resembles the heather bee of northwestern Europe and not the present English bee. This fact affords evidence against a common assumption that the black bee was introduced here by early colonists from England, an assumption not confirmed by any historical record.

the blacks. Where honey superabounds, the blacks do well enough; but when it is scarce and can be got only by unusual energy, then the superiority of the Italians is very manifest. (17) While black bees assert no claims to food offered them away from their hives, Italians will often try to prevent other bees from getting any of it, as stoutly as though it were a part of their own private stores. (18) Italians will utilize largely any wax which they find away from their hives, while the black bees take no notice of it. (19) Black bees, when examined by artificial light, are more inclined than Italians to fly from their combs. (20) Black bees have a much stronger attachment than the Italians to the spot where their hive once stood. Dzierzon, when he had only blacks, found it highly desirable to have two apiaries far enough apart to enable him to secure enough bees for his artificial swarms and nuclei; while many of the methods given in my work, and which cost me so much time in observations and experiments, aimed to secure the same results from a single apiary. Those who have had no experience with blacks, have but little idea what a task it was, in many of the most important operations with them, to get a sufficient number of bees that would stay in any new location. (21) When the union of blacks from different colonies is attempted, they are far more likely to quarrel than are Italians. (22) While I do not claim to have given all the points of difference between these two races of bees, I have been the more particular, because of the conviction that so few are now living of the old generation of beekeepers who have had a sufficiently long and large experience to be able to give the facts on the subject. Of one thing I am sure,—that the Italians are in greatest favor with those who are best acquainted with the striking points of difference between them and the blacks, and that the use of movable frames, with all the manipulations which follow in their wake, have set a seal of condemnation upon black bees which can never be removed.

IX

Not to the Swift

"I returned, and saw under the sun, that the race is not to the swift, nor the battle to the strong, neither yet bread to the wise, nor yet riches to men of understanding, nor yet favour to men of skill; but time and chance happeneth to them all."—Ecclesiastes 9. 11.

§1. INVENTOR BALKED OF PROFIT

LANGSTROTH'S patent in the improved hive brought him little or no material benefit, although the general use of the Langstroth principle, a frame isolated by the bee space, transformed the processes of bee culture in the United States within his lifetime. There were several reasons for his misfortune. He was not a good business man to begin with. His nervous infirmity disabled him for months on end. When he obtained his patent he had no way of arranging for the exclusive manufacture and sale of such an article as the movable-frame hive because there was not, then or for years afterwards, any mass production of supplies for beekeepers or any ready medium of direct advertising to them. At any time or in any circumstances it would have been difficult to monopolize such an article, which a village carpenter, once given the principle of the thing, could make cheaply. He was robbed of his rights by wholesale infringement of his patent. And the story of efforts to prevent the robbery is a chapter of accidents.

Within a dozen years after the issue of his patent the sale of hives which infringed it became so extensive that he took steps to repair his defenses. In 1863 he asked for and obtained a reissue as a means of supplementing his original application with more particular evidence to substantiate his title.

§2. OPERATIONS UNDER THE PATENT

The patent was to expire in 1866, but it might be and was in fact extended for seven years. When the time came to apply for the extension Langstroth was too ill to attend to the matter himself and his wife prepared an affidavit that was called for.[1] The document was written under date of Oxford, Ohio, July 25, 1866, addressed to the Commissioner of Patents, and signed by Anne T. Langstroth and L. L. Langstroth. It appears to be a carefully composed response to the Patent Office's request, formal and customary on such an occasion, for certain information "particular and in detail." It reveals so much of the history of the inventor's difficulties in promoting his business venture that it is worth printing here in full. It follows:

I address you on behalf of my husband, Rev. L. L. Langstroth, who has recently filed an application for an extension

[1] A photostatic copy of this affidavit has been added to the Langstroth memorial collection in the Cornell beekeeping library through the courtesy of Parker Dodge, attorney-at-law of Washington, D. C.

of his Patent on Bee-Hives. As the state of his health precludes the possibility of his making the necessary statement with regard to the pecuniary profit he has derived from the Patent, and the means he has used to publish his invention, it devolves on me to do it for him.

You ought to be made acquainted with the embarrassments arising from the effect of disease upon his mental as well as physical system, as these have operated largely to prevent his deriving any profit from his invention, at all proportioned to its value, or to his "time and expenses in discovering and perfecting it."

No one but myself can give you any adequate idea of these embarrassments, and I trust I shall not weary your patience, Sir, if in making this statement, I go back to the time of our marriage, now nearly thirty years ago.

Mr. Langstroth had then just completed his Collegiate and Theological education, and was a settled pastor. Before the first year of his settlement had expired, he had frequent attacks of distress in the head, which entirely incapacitated him, for the time, for any mental effort. He had suffered in the same way, to some extent, during his course of study; but these attacks now becoming more severe and frequent, he was compelled to ask dismission from his people, in less than three years from the time he became their pastor. For several years succeeding, he was engaged in teaching; still the difficulty in the head continued, but his health somewhat improved, so that he felt he might return to the active duties of his profession as a minister of the Gospel. The result was similar to the former. Frequent and still more protracted seasons of severe distress, attended with great mental depression, and an almost entire prostration of physical strength wholly unfitting him for study, he was compelled, after a settlement of a little more than four years, to return to teaching. By this time, however, his disease had assumed a more aggravated form, so that he was unable to discharge the duties even of a teacher, and it became necessary to give up a sedentary life, and devote his time and energies to some other pursuit.

He had for several years kept bees, the care of which afforded him pleasant recreation from his professional duties. A remarkable interest in Natural History, manifested from early childhood, was thus gratified, and his desire to understand, by personal observation, the mysteries of the hive, had induced him to use a Huber hive, and afterwards one made of glass, of large dimensions, but containing only a single comb, so that he might see all that was going on within. Not finding any kind of hive suited to his wishes for practical purposes, he made one improvement after another, until he devised the hive, patented in 1852.

Since that time, he has devoted himself mainly to apiarian pursuits, but in these, he has struggled against difficulties which would have entirely discouraged one possessed of less energy and perseverance; for though while suffering from these attacks, he is much depressed and almost hopeless with reference to future success, when well he is full of courage and enthusiasm in his pursuits. For the last seven or eight years he has been laid aside almost entirely from all physical and mental labor, about one half of each year, and the necessity of this document being prepared by another than himself, will be apparent when I state that he has been suffering so greatly for the last seven months, that during all this time he has not written a letter either of business or of friendship.

After obtaining his patent, Mr. Langstroth wrote his treatise on the "Hive and Honey-Bee," and has since prepared two revised editions. In this work he was greatly hindered by the difficulty in his head, being often obliged to suspend his labors for many months; after a part of the second edition had been printed, he was compelled to intermit his work upon the remainder for about six months. A new edition, embodying his later discoveries in Apiarian science, and his experience with the Italian bees, he has been, for some time, anxious to prepare, but has as yet been barely able to commence it; the distress in the head returning, and almost overpowering him, as soon as he gives himself to mental application.

I will only add to this narration, that the intervals of relief from this state of suffering have been largely devoted not only to efforts to introduce the hive to public use, but to observations and experiments tending to perfect it and the system of management connected therewith.

I have been thus particular, Sir, in my account of the disappointments and discouragements experienced by Mr. Langstroth during this long period, in consequence of ill health, believing that it would show "sufficient reason" why a statement more "particular and in detail" cannot be given. In considering the statement of receipts and expenditures about to be submitted, it should be borne in mind:

That Mr. Langstroth's education and habits were of such a nature as to seriously disqualify him for the business of vending a Patent Right;

That the invention, to be valuable to the masses, required a thorough education of beekeepers in the new system of culture which it inaugurated,—a work of years of toil with the pen and in person, and only within the last few years, in some measure accomplished;

That Mr. Langstroth's ill health has prevented him from devoting the time and attention to the introduction of the Patent which he would otherwise have given to it; while his pecuniary circumstances have been such as to compel him to make sales at prices far below what he regarded their real value; and to prevent him from investing the sums necessary to canvass the field, as he might otherwise have done;

That one-half the interest in his whole invention passed out of his hands, as will be explained hereafter, for a very small consideration, and that the most available territory of his remaining half interest, was sold for a small sum, but little even of which was paid;

That of the remainder of his half interest, the Southern States constituted the larger part, and that, as during the war no sales of territory could be made in those States, a large part of the allotted time for the Patent, has been lost, with reference to them;

That owing to Mr. Langstroth's ill health (our son, now a partner with his father, being then engaged in other business) no books were kept, nor any regular accounts of receipts and expenses, until the last three years.

I am now to present a statement of the receipts and expenses as estimated by Mr. Langstroth, according to the best of his ability:

To Dr. Joseph Beals, of Greenfield, Massachusetts, a capable business man, in order to procure means for constructing hives and introducing the invention, was given a half interest in the whole Patent.

To R. C. Otis, of Kenosha, Wisconsin, was sold the Patent Right for the Western States and Territories. Mr. Langstroth received for his half interest in this sale, not over $1800, as payment of a large part of the nominal price was made in notes of other parties, to whom Mr. Otis had sold portions of his territory, from which, though supposed to be good, no money was ever realized; said notes having been protested and afterward surrendered by Mr. Langstroth to the parties giving them, in exchange for the territorial rights they had purchased of Mr. Otis.

In the $1800 above specified is included the sum of $1200, received by Mr. Langstroth from Richard Colvin, of Baltimore, Md., for territory in Ohio, sold to Colvin; said territory having been reconveyed by Mr. Otis to Mr. Langstroth in exchange for Mr. Otis's note of $1000 given to Mr. Langstroth at the time Mr. Otis made his original purchase.

To other parties, territorial rights have been sold, yielding, in the aggregate, a sum not exceeding $1500. To various persons, individual or farm rights have been sold, yielding in all the fourteen years not more than $1200. From sales of hives, during the whole period, the profits arising have not exceeded $800.

The expenses attending the business transactions connected with the invention can only be estimated; and for hives, models, drawings, traveling expenses, legal fees, etc., will nearly, if not quite, reach the sum of $2000; and this, without taking into account the value of the time donated by Mr. Langstroth to the business.

As to "the ascertained value of the invention," required to be stated by the applicant, it is somewhat difficult to make the statement on such an article as a beehive. Mr. Langstroth, however, believes that he correctly estimates its value to the Bee-keeper at from two to three dollars on every hive of his device.

In the statement of receipts and expenditures which the above affidavit includes there is only a vague account of the consideration that was agreed upon in return for the half-interest assigned to Dr. Beals of Greenfield. Langstroth wrote of that transaction just as indefinitely in his Reminiscences, many years later, but there he said enough to show that Dr. Beals had come to his help, made an acceptable offer, and executed it faithfully.

"When I determined to apply for a patent," he recalled, "as I had neither the money nor the business qualifications needed for its profitable introduction, I thought myself fortunate in being able to secure the services of a firm which had been quite successful in selling patents; but business reverses prevented them from carrying out our agreement. While [I was] writing my book, Dr. Joseph Beals, one of my former Green-field parishioners, offered, for an interest in the invention, to furnish means for the manufacture of the hives, and for establishing an apiary [to replace the one which had been sold in West Philadelphia the year before]. Dr. Beals was a very able dentist, but without any experience in patent matters or in beekeeping, while I was frequently prostrated by the old head trouble. Although much was done to introduce the movable-comb hive to the public, we met with no adequate pecuniary success; and after some years we closed up our business, without any abatement of the old friendship, the doctor having lost much time but no money by his venture." [2]

The partnership with Dr. Beals may have been dissolved before Langstroth left Greenfield and took his family to their new home at Oxford in 1858. During the next seven or eight years, the above affidavit declares, he was "laid aside almost entirely from all physical and mental labor" for about half of each year. His only son, James Tucker Langstroth, who came of age in 1858, was taken into partnership and proved a strong ally, but he enlisted in the Union army early in the Civil War, obtained a commission, and served for several years. After his discharge he suffered from tuberculosis contracted in camp and died in 1870.

§3. THE CASE OF OTIS VS. KING

Otis, the Wisconsin man who had bought from Langstroth and Beals the patent right for the sale of hives in

2 *Gleanings in Bee Culture*, xxi, 206–7.

"the Western States and Territories," had come, in the course of events, to have as much interest as anybody in keeping the patent alive and in preventing infringement. His influence can be felt in the urgency of the applications that were made to the Patent Office for the reissue in 1863 and for the extension in 1866. He had built up a large business, partly through direct sales and partly by disposing of territorial rights within his domain. He had a valuable property there if trespassers could be kept off. The West had been filling up, and now an even greater tide of settlers was flooding in and claiming agricultural lands under the Homestead Act of 1862.[3] It was a growing market for the beehive, especially for the new one with the easily movable frames, whose singular worth was becoming generally known. But Otis's agents reported unfair competition. Other hives with the movable frames were sold under various subterfuges. Some had been patented which made no claim to the Langstroth frame but actually included it.[4]

An artful competitor was Homer A. King, who was engaged in business with an office at 240 Broadway, New York. He was making trouble for both Langstroth and Otis. Eventually a suit, in Otis's name, was brought against him for infringement of patent. Notice of inten-

[3] For example, the aggregate population of three of the States within Otis's territory, those of Michigan, Illinois, and Wisconsin, increased between 1850 and 1870 from one and one-half millions to four and one-half millions.

[4] For example: While the first national convention of American beekeepers was in session at Indianapolis in 1870, three years before the Langstroth patent was to expire, time was given, five minutes to each exhibitor, for the showing of "eighteen movable comb hives"! *American Bee Journal*, vi, 171.

tion to prosecute was given by Langstroth in a statement dated at Oxford, Ohio, March, 1871, and published in the *American Bee Journal* (vol. vi, p. 219) under the head "H. A. King's and L. L. Langstroth's Patents." The statement is technical; because the authentic inventor and patentee is explaining a technicality by which he was tricked:

In the spring of 1867, Mr. H. A. King entered into an arrangement with L. L. Langstroth & Son, by which he agreed to pay a certain sum on all sales of hives, rights and territory subject to his patents, when such sales were made in territory still owned by Langstroth. The agreement confined him to the use of certain *slots* in the top bars of his frames, for admitting bees to top boxes, as shown in a model deposited with Langstroth & Son—that is, he was allowed to use the Langstroth frames with tops partially separated, and no other patented features of that invention.

On September 8th, 1868, Mr. King took out a patent under which he no longer uses the *slots* or notches by which the tops of his frames were partially separated, but substitutes *mortices* for them in the tops of the frames, thus allowing those tops to fit *closely* together throughout *all* their length. To inquiries frequently made whether we considered those mortices an infringement of our patent, if used without proper license, we replied in substance that we did not. It was obvious to us that this mortice enabled Mr. King to use an important feature in my invention, and one very fully set forth, both in the original patent and in the reissue, viz., the allowing bees to pass above the frames into supers so that the honey might be obtained in the most beautiful and salable form, and be safely removed from the hive even by timid and inexperienced persons—a thing never even *contemplated* in any movable-frame hive before mine. Still it seemed to me that it did not conflict with the *wording* of my claims, and that therefore I could not prevent its use without another reissue and better wording of my claims; and as such reissue would have relieved all parties from liability for any previous infringement, we thought it best to acquiesce in its use.

In the spring of 1870, only a few days before the death of my son, Mr. King notified us that as he had not for some time used the notches or slots for which he agreed to pay us a percentage on all sales in our territory, he must decline paying anything more under that agreement.

Having now recovered my health so much as to be able to examine his patent more thoroughly, and having taken the ablest legal advice to be procured, I am satisfied that Mr. King's mortices will be pronounced by the Courts to be "a mere colorable evasion" and therefore substantial infringement on my rights. Having already, in a personal interview, informed Mr. King of the view I now take of the matter, justice to him and to those parties intending to purchase under him, supposing that his hive is confessedly no infringement upon mine, renders it proper that I should make this public statement. Those parties also who have purchased under my patent, and who have been damaged in their pecuniary interests by an opinion given by me without proper legal advice, have the right to demand that I should take the earliest opportunity to state that I regard the use of Mr. King's mortices or any equivalent device for the purpose of passing bees above the frames into boxes, to be an infringement upon my patent, unless licensed to be used by the owners of said patent; and that the earliest possible steps will be taken to have the matter decided by the United States Courts.

Appended to the above statement in the *American Bee Journal*, and bearing the same date, March, 1871, was another communication signed by Langstroth, addressed "To the Beekeepers of the United States." It was an account of the trouble he had been put to by persistent efforts to impugn the validity of his patent. His application for its extension in 1866, he said, had been "hotly contested." The testimony of some of his opponents, on record in the Patent Office, would, if made public, he declared, "have consigned some of them to infamy" and might, if pressed home, "have placed others in the penitentiary." He quoted "an extract from a communication published in the *Prairie Farmer*, in October, 1866," as follows:

Prior to the reissue of Mr. Langstroth's patent in 1863, the opposition had relied on the idea that his patent was anticipated by foreign inventions; but at the time of the reissue, and on the hearing now [in 1866, when the extension was applied for], *Mr. Langstroth himself furnished and laid before the office every work having any bearing on the subject, both foreign and native, nearly thirty in number,* embracing some very rare works —one being the only copy existing in this country. On the recent hearing, they abandoned, wholly, the idea of its being anticipated by any foreign invention, and relied on the effort to prove a prior invention in this country—*no less than four of them* swearing that they had invented or used the same thing prior to Langstroth! But these parties had done what Job so fervently desired his enemies to do—they had each of them "written a book"!—and those books—if there had been no other testimony—were sufficient to decide the case against them. There probably has never been a case in the office in which there was so much of fraud and perjury as was furnished on the part of the opposition in this case; and it is no wonder that both the Examiner and Commissioner came to the conclusion that the testimony was *"not worth consideration."*

Concluding this second statement, Langstroth said he hoped that every legitimate weapon which could be used to break down his patent would be brought forward in the case now about to come into court. He invited "all the beekeepers of the United States, and all anywhere else" who might see this appeal, "to send to H. A. King and Co., 240 Broadway, New York," against whom, he said, suit had been brought for infringing on his patent, any evidence which might tend to weaken, limit, or invalidate his claim. He concluded: "I stand upon what I believe to be my rights. If I have none, but am unfortunate enough to be the honest *original* inventor, who, to his surprise and sorrow, finds that he was not the *first* inventor, the sooner I know this, the better; that I may at once cease from claiming what would then belong to the public, and not to me." [5]

The case of Otis vs. King never came to a decision in court. A series of fatalities prevented. An account of what happened is found in a letter written by Langstroth in February, 1875, and published as a supplement to the April number of *Gleanings in Bee Culture*, four years after his public announcement of the suit. His too brief report of the catastrophe makes a fitting anticlimax:

". . . Since the last of June 1873, I have been laid aside from business of all kinds. . . . Two years ago I was straining every nerve to have the case of Otis against King brought to an issue. That eminent counsellor S. S. Fisher, after seeing all that the defense could say for their case, was confident that the claims of my

[5] *American Bee Journal*, vi, 219–20.

patent would not be invalidated. The day was set for the hearing [before Judge Swayne in Cincinnati [6]]; but before the cross-examination on my own sworn statement could be completed I was prostrated in mind and body by my old complaint, and everything came to a stand. Since then Col. Fisher has died; and Mr. Otis, after being some time an inmate of an insane asylum, died there, and of course the suit came to an end."

He went on in this letter to say that his relatives had known, and he himself had told King, that in aiding Otis he did not intend to appeal to the law to maintain his own rights against infringers. His settled and declared intention was to leave all infringers, in the large territory which he then owned, to act as their own consciences might dictate and pay him a license fee or not, even though the law allowed him seven years after the patent's expiration for the collection of damages. "I have felt for years," he wrote, "that from the many conflicting and, as I believe, infringing hives which have come into use my relations to the beekeeping community had become misunderstood by many who were ignorant of the facts. I have unceasingly grieved to find myself, in my old age, in such unpleasant antagonism to many with whom I sought to maintain only friendly relations."

6 "Private Letter," *American Bee Journal*, viii, 263–4 (1873).

§4. GOOD MEN VINDICATE
LANGSTROTH

Langstroth's public announcement of the suit for in-fringement had stirred up a controversy in the journals. Persons with mercantile interests at stake took advantage of the delay in the prosecution of the case and presumed to try it in the papers in any but a judicial manner. On his part King had the use of a house organ of his own which he called the *Beekeepers' Journal,* having chosen a title that seemed to some beekeepers to invite confusion with Samuel Wagner's established and reputable *American Bee Journal.* Insinuations and half-truths were printed which put Langstroth in a false light, calumnies which outlasted his patent rights and reappeared to trouble him years after anybody could have had a mercenary motive for defaming him. But champions came to his side, men of character and standing who knew how to declare the worth of his work and were eager to declare it for truth's sake. They proclaimed their knowledge that his invention was unique and of singular importance. Some of these statements, extracted from bee journals of the '70's and '80's, are printed below.

A champion as well-armed and sturdy as any was Samuel Wagner, but he died early in 1872. While he lived he put his *Journal* at his friend's service. In an article on the history of efforts to improve the hive he wrote: "When Langstroth took up this subject, he knew well what Huber had done and wherein he had failed—failing, possibly, only because he aimed at nothing more than constructing an observatory hive suitable for his purpose. Mr. Langstroth's object was other and higher. He aimed at making frames movable, interchangeable, and practically serviceable, in bee culture. *Nobody, before Mr. Langstroth, ever succeeded in devising a mode of making and using a movable frame that was of any practical value in bee culture."* (The italics are Wagner's.)

John Phin, professor of agriculture in Pennsylvania State College and author of a *Dictionary of Practical Apiculture,* wrote:

I have rarely been more deeply interested in any article than in that which appears in the *American Bee Journal* in relation to the hive invented by Mr. Langstroth. That he should now, after all his years of faithful labor, be poorer than when he first began the work, does not speak well for the honesty of American beekeepers. One thing that strikes me most forcibly is the seeming ignorance of the writers in the bee journals in regard to the points for which we are indebted to Mr. Langstroth. I have examined the subject with a great deal of care, and am fully convinced that every man who uses comb frames constructed and arranged as in the Langstroth hive is using that which does not belong to him. And yet we find men discussing the merits of the different hives and contrasting them with Langstroth's, while at the same time these other hives owe their most valuable feature to Langstroth's ingenuity.

I know that difference of shape, material, and interior arrangement may properly be the subject of discussion and comparison with Langstroth's form; but in this discussion we are too apt to forget that to Mr. Langstroth we owe that which is far more important than any mere form arrangement. Take away from beekeepers the knowledge of the movable frame, and where should we be? I have no pecuniary interest in the matter. I am not even personally acquainted with Langstroth. I speak in simple justice to a worthy man, to whom we all owe not only a deep debt of gratitude, but of cash. Can we not afford to be honest?

The following bit of testimony is extracted from an article written by Dr. Charles C. Miller [7] of Marengo, Illinois, who for many years was recognized as a leader by a host of beekeepers in North America:

It is well known that among the many hives in use, no other make is so popular as the Langstroth; but it may not be so well known that in a very important sense, every hive in use among intelligent beekeepers *is* a Langstroth; that is, it contains the most important feature of the Langstroth hive—the movable frame. Those who have entered the field of apiculture within a few years may faintly imagine, but can hardly realize, what beekeeping would be today if throughout the world, in every beehive, the combs should suddenly become immovable, fixed, never to be taken out again except when they were cut out. Yet exactly that condition of affairs existed through all the centuries of beekeeping, up to the time when, to take out every comb and return it again to the hive without injury to the colony was made possible by the genius of Mr. Langstroth. It is no small compliment to his far-seeing inventive powers that although improvements, so-called, upon his hive have been made by the hundred, one after another these have dropped into oblivion, until the thousands of hives in use among the best

[7] The University of Wisconsin's excellent Miller Memorial Beekeeping Library is named for him. Its endowment is a fund to which beekeepers of all parts of the country contributed after his death in 1920. The formal gift of the endowment fund at Madison was followed by a journey of the principals to Marengo and the dedication of a memorial tablet in the church which Dr. Miller had regularly attended. The Cornell beekeeping library has numerous volumes of bee journals which belonged to him.

beekeepers scarcely vary, if varying at all in principle, from the Langstroth hive as first sent out.

G. W. Demaree of Christiansburg, Kentucky, was the discoverer of a method of swarm control which is now used, with numerous modifications, by most American producers of extracted honey. He wrote this statement:

While watching the drift and current of bee literature of the past, I have noticed that the subject of beehives and frames has periodically come up for discussion. There is about the Langstroth frame a history the most remarkable of anything connected with the inventions of many years, defying the genius of the American people. Thousands of efforts have been made to supersede it with something better—with a better frame for all purposes; but not even a large minority of beekeepers at any time have been induced to accept anything as superior to the Langstroth frame.

One persistent argument against the priority or the importance of Langstroth's invention was that Dzierzon had anticipated him in the use of movable frames. Germans in particular were disposed to take up the cudgels for Dzierzon. Although he was a Pole, he was a native of a province which Prussia had seized in the eighteenth-century partition of Poland and had imposed the German language upon. The fame of his contributions to bee culture, written as they were in German, gave his methods great favor in the German-speaking countries. In Selma, Texas, there was a progressive beekeeper, an immigrant from Germany, who read German books and periodicals dealing with the industry and wrote articles for American bee journals which he signed L. Stechelhausen. Following is an extract from an article of his which was published in 1888:

Dzierzon never used a frame, but comb-bars only. The bees build the combs to this bar and to the walls of the hive. His hives were at first different styles, because he adjusted the bars into his old hives. So he used about fifty years ago shallow boxes, forming the so-called Christ's magazine hive, with these comb-bars. This hive was quite similar to Heddon's hive, and was manipulated from above. To take out any comb it was necessary to cut them loose from the hive walls. It is easily seen that this is somewhat difficult if managed from above. This is the reason why Dzierzon had abandoned this kind of hive, and adopted hives manipulated from the side or rear.

It has been proved, many times, that the first hanging, movable-frame hive was invented by Langstroth in 1851. A short time later, Baron von Berlepsch, of Germany, changed Dzierzon's bars to frames. It is true that he made his frame quite independently of Langstroth. Berlepsch did not change the manipulation from the sides and these hives are still in use in Germany. Dzierzon still recommends using the comb-bar; and because nobody in Germany or elsewhere is on his side in this respect, he conceded, finally, that frames may be used with advantage in the surplus chamber for extracting, but none in the brood-chamber.

Within the last few years I have written articles for different German papers, in which I explained the advantages of our American hives and management; this caused Dzierzon to advise us not to use a hive with frames to be manipulated from above, because by taking out the first frame, the bees would be rolled and killed; and other beekeepers had other objections.

Nevertheless, our hive system is gaining more friends in Germany. No one can honor Dr. Dzierzon more than I do, but to call him the inventor of a frame hive is just amusing, when he used every occasion to speak and write *against* frames.

Another type of hive that figured in the controversy was the device of Baron August von Berlepsch. Some persons asserted that Berlepsch was the actual inventor of the movable frame. That assertion was confuted by Charles Dadant,[8] who wrote, in 1885:

While Dzierzon and Berlepsch in Germany were working to find a movable-comb hive, Munn in England, Debeauvoys in France, and Langstroth in the United States labored toward the same end, and it is Mr. Langstroth who became the winner in the race, as it has been proved beyond any question of doubt that he applied for his patent about six months before Berlepsch had invented his hive. In regard to his own invention, Berlepsch wrote: "Until 1851 I had the misfortune of using movable-comb hives so miserable that my work was tiresome or delayed. . . . At last, after seven years of silent work, I came to the front in 1853 and 1854 with my letters on beekeeping, having the solid ground under my feet."

The inventions of Munn and Debeauvoys are already forgotten, and the differences between the Langstroth and the Berlepsch hives are so manifest that nobody could possibly think the one suggested the other; for the only point of resemblance is the space between the frames and the sides of the hive, a space indispensable to the removing of the frames, to which Dzierzon has always been, and is even now opposed.

The frame hives of Berlepsch, like the bar hives of Dzierzon, have their combs parallel to the entrance and open at the rear by doors, like cupboards. They were adopted as standard by bee-

8 For an account of his work, see above, Chapter VI.

keepers in Germany and Italy, but some beekeepers began to try hives opening at the top, as invented by Langstroth, and the comparison proves so much in favor of the latter that I hope sooner or later to see the German standard yield to the American, as I have long prophesied that it would.

X

A Chapter of Difficulties

A FEW circumstances and events of Langstroth's life deserve fuller treatment than could be given them in a continuous narrative and are reserved for this chapter. They are (1) his nervous malady and its consequences, (2) his interrupted but nevertheless devoted service of the Church, (3) what the Civil War meant to him, (4) the help that came to him in his need from kindly relatives and particularly from three brothers-in-law, and (5) an act of his in his late seventies that puzzled and hurt some of his best friends, his unqualified recommendation of the Heddon hive.

§ 1. HIS BAFFLING INFIRMITY

Attacks of a recurrent malady prostrated Langstroth in mind and body for periods which—to judge from evidence found in his letters—amounted to half of his adult life. Only an extraordinary power of mental application when he was well could have enabled him to accomplish all the work he did in the course of those years. He proved that he could work fast and to good effect when, for example, he wrote his masterpiece, a book of nearly 400 pages, in a single winter.

He never found a physician who could get at the root of his trouble or could even give him a helpful hint of the nature of the disease. He left a written account of his experience which suggests that his was a case for a psychiatrist—if he could have found one. The account becomes in part a soliloquy which one shrinks from overhearing:

For many years [he wrote] I have suffered from what I have been wont to call my "head trouble," which not only unfits me for mental exertion, but also disqualifies me for enjoying almost anything personal to myself. While under its full power, the things in which I usually take the greatest pleasure are the very ones which distress me most. I not only lose all interest in bees, but prefer to sit on that side of the house where I can neither see nor hear them. Gladly, if at all convenient, would I have my library of bee books hidden from my sight; and often I have been so morbid that even the sight of a big letter *B* would painfully affect me. At such times, fearful of losing my reason if I allowed my mind to prey upon itself, I have resorted to almost constant reading to divert my thoughts. The greatest objection to this is that it not only fails to interest me when I am most unwell, but, by association of ideas, too often deepens my distress. To use the words of the old poet Herbert [*The Temple*, "Affliction" IV, No. 65],—

> "My thoughts are all a case of knives,
>   Wounding my heart
>   With scatter'd smart."

I have, therefore, for years, read less and less, and occupied my time mainly with chess, which is too impersonal to suggest the melancholy ideas which so often torment me when reading.

As soon as I awake I try, by chess problems the most intricate that I can find or invent, to forestall the approach of gloomy thoughts, continuing to play as though a fortune could be made by it, or as if I were playing for my very life; and often, during the larger part of the night, my brain seems to be incessantly moving and supervising the pieces on the chessboard.

Methinks I hear someone exclaim, Can this be the condition of a minister of the gospel of Christ? Ought not the blessed promise of God's word always to enable him to attain, in some measure at least, to the apostle's experience when he said, "Now the God of hope fill you with all joy and peace in believing, that ye may abound in hope, through the power of the Holy Ghost"? No! No! God has not promised to overrule his natural laws by constant miraculous interposition. Can you give a wholesome appetite for food to a person intensely nauseated, by merely showing it to him and inviting him to sit down and partake of it? It is a great help, doubtless, even under the most depressing circumstances, to know that God is good, and to hope that in due time the dark side of the picture will be turned from us, and its bright one again be displayed. This hope often sustains us when otherwise we might be utterly cast down.

Read the Forty-second Psalm if you doubt what I affirm. "Hope thou in God; for I shall yet praise him for the help of his countenance. O my God, my soul is cast down within me. . . . All thy waves and thy billows are gone over me. . . . Why art thou cast down, O my soul, and why art thou disquieted within me? Hope thou in God, for I shall yet praise him, who is the health of my countenance, and my God." Not now, oh, not now! but I shall yet praise him.

It is evident that the Psalms which express our strongest emotions could have been born only out of personal experience; some, "when gladness wings our favorite hours"; others, when we are almost disposed to repeat that anguished cry of our Saviour, "My God! my God! why hast thou forsaken me?" Only thus originating could they have lived in the memory of man for so many ages. As in water face answereth unto face, so the heart of man, and I earnestly hope that some afflicted brother or sister who has been crying out, "All thy waves and thy billows are gone over me," may be helped by realizing that the great Father of our spirits, who pitieth his children, who knoweth their frame, and who remembereth that they are dust, has caused special Psalms to be written, even for them. . . .

Never was I able to get the better of these incipient attacks, no matter by what method I tried. It might be of longer or shorter duration before it prostrated me; but it always had but one issue. Struggle as I would, fight as I could against it, my condition was that of the man lost in the quicksands, so vividly described by Victor Hugo. Walking carelessly over its treacherous surface, he first notices that his freedom of movement is somewhat impaired; but he thinks little of this until he finds it more difficult to lift his feet. Alarmed at last, he vainly tries to escape to the firmer land, only to find that each step he takes sinks him deeper and deeper. . . .

The first light thrown upon my case was by a German physician who told me that my trouble was due to blind piles; but he failed to cure me.

I shall never forget the remark of an electric physician, who, in 1853, while passing his hand over my neck, exclaimed, "How can a man with the flesh over his spine in such a rigid condition be otherwise than miserable!" It was the first time that my attention had been called to the abnormal congestion of the flesh over the whole length of my spinal column. "You will be happy," said he, "as soon as I relieve you of this congestive condition." He worked upon my spinal column for several hours, much as they do in massage treatment, till at last he effected what seemed at the time to be a cure. But he died before I could avail myself of another treatment.

So intimate is the connection between this rigidity and my mental depression that they are never dissociated; but in vain I have called the attention of able physicians to this feature of my case. When it began to develop they never succeeded in arresting it. . . .

I ought to correct what I said about there never being but one issue to an attack after its incipient stages had clearly developed. In the fall of 1853 I was as much depressed as I had ever been, when, by the kindness of friends, I was able to visit a brother who was residing in Matamoras, Mexico. While traveling by steamboat, railroad, and stagecoach to New Orleans—a journey which then occupied more than a week—I recovered entirely before I reached that city, and had an unusually long interval of complete relief. Also on another occasion when greatly despondent, I was summoned, at the expense of one of the parties, as a witness in a suit at law, which had been brought against him for an alleged infringement on the right of another patentee. The entire change of scene, with its many diversions, completely cured me. But for these instances, I might naturally infer that time was the only remedial agent, and that the disease could never be arrested, but must always run its usual course.

Among the mistakes of my life, I count this to be one of the greatest, that, instead of seeking an entire change as soon as I began to feel the approach of another attack, I have usually refused to admit the possibility of succumbing to it, and have struggled against it until no power of will was left for further conflict. Those who know how large a portion of my life I have lost by this disease will not be surprised at my unwillingness to quit my work, when to give it up often meant to forego opportunities never to be recalled. Besides all this, I have usually been so straitened for means that it has been very difficult for me to give up my necessary avocation for a change of scene.

No doubt there are some who would blame me for spending so much time, when under the power of melancholy, in playing chess, even though I tempted nobody else to waste time upon it. But I most devoutly believe that in fighting such a malady the end fully justifies all means which are not in themselves immoral. It would be well, if it were plainly understood and more fully realized, that, by dwelling too long upon painful subjects, we may at last lose mental control and become absolutely insane. There is no doubt that many who have strong hereditary tendencies that way may, by wise foresight and strong effort, counteract them.

§2. HIS DEVOTION TO THE CHURCH

Langstroth's physical unfitness for the regular practice of the Christian ministry was a cause of sorrow to him all his life. He never willingly gave the ministry up. Until his death, which occurred in a church and just as he was beginning the delivery of a sermon, he was often responding to a request or invitation to supply a pulpit. His wearing of the clerical collar—not the usual habit of the Congregational minister—seems to have been for him a tonic symbol of devotion to his calling, for he was not given to affectations.

His early pastorates at Andover and Greenfield were the only ones he ever held. The first, of two years, 1836–38, he resigned with regret. The second, which was offered him after he had preached the sermons for two years, he accepted hopefully and held for another five, 1843–48. From then until 1852, while he lived in Philadelphia, though he had no pastoral charge, he was in one pulpit or another, as he remembered, on more than half of the Sundays. After his return to New England in 1852 he supplied the pulpit of the Congregational Church in Colrain, near Greenfield, the larger part of the time till the fall of 1857.[1] Within that span of about twenty years he must have given much of his energy and time to the Church.

----

[1] Reminiscences, *Gleanings in Bee Culture*, xxi, 206–7.

At Oxford, Ohio, where he had his longest continuous residence, 1858–87, and at Dayton, until his death in 1895, he was no less eager to associate with fellow clergymen and to give them whatever professional help he could. In Dayton his closest associations were with Presbyterians. He attended the Park Presbyterian Church and followed its pastor, the Reverend W. F. McCauley, to the Wayne Avenue United Presbyterian Church, which had grown out of a Sunday school established by his son-in-law, H. C. Cowan, with whom he made his home at that time. He was a friend of the Young Men's Christian Association, attended its Sunday afternoon meetings, and sometimes spoke at them. Among his ministerial associates he came to be known affectionately as "the patriarch" of the Dayton presbytery.

§3. SLAVERY AND THE WAR

In the Civil War Langstroth's loyalty to the cause of the Union was unhesitating. He encouraged his only son to volunteer for military service and made no complaint when his loss of the young man's help at home left him to carry a burden of work alone.

What the war meant to him, at least in retrospect, can be read in his Reminiscences, written more than twenty-five years afterwards.[2] There he has a great deal to say about the slavery question. No doubt that question gave the Civil War for him, as for many righteous

2 *Gleanings in Bee Culture*, xx, 912–14; xxi, 8–9.

people of the North, a moral issue, not so generally to be found in political principles or economic ideas. But in him there was inbred a special cause for feeling about it. As a boy he had heard many stories about Negro slavery from his mother, whose family were planters on the Eastern Shore of Maryland. Her mother, Elizabeth (Lorraine) Dunn (see Chapter I), had inherited a large number of slaves but had freed them. "When she was a rich and fashionable widow," her grandson related, "the converting grace of God came to her through the ministrations of the Methodists, and she joined their church, although to connect herself with them was almost to lose caste with many of her associates. . . . Long before she died, my grandmother liberated all her slaves. . . . The hated institution! If possible, its curses rested more heavily upon the whites than the blacks; and it is often difficult for me to realize that I have lived to see it overthrown."

In the later years of his residence at Yale, from 1831 till 1836, agitators in the North took up the slavery question. It was in 1831 that William Lloyd Garrison began publishing *The Liberator* and helped to organize the New England anti-slavery society. In this excited controversy Yale College was a sensitive spot because it had a good many students from the South. Incidents of that time were impressed on Langstroth's memory. He recalled that President Day, though he hoped to see slavery done away with, did not approve of anti-slavery societies in the North because he feared that they would thwart their purpose by exasperating Southern men. He relates that he once heard Day say, in effect: "If a door is partly open, and you for some reason wish to have it opened wider, it would not be wise to use such irritating language as could only end in having the door slammed in your face."

At New Haven he heard Garrison, in a public address, interrupted by angry shouts: "Many Southern students were present, and great offense was given them by the opprobrious epithets which [Garrison] so vehemently bestowed upon all slaveholders. I could easily see that their sense of justice was often violated, and that they could not fail to be provoked, by his strong denunciations of their Christian fathers and mothers. I suppose it would, at that time, have been an inconceivable idea to Mr. Garrison that 'men-stealers,' as he called all slaveholders, could possibly be real Christians. The cries, 'It is a lie!' 'You are a liar!' were hurled at him by those who truly believed that he deserved such epithets, and the meeting broke up, I believe, in confusion."

"It may well be doubted," Langstroth wrote, "whether the steps which led to the overthrow of slavery could ever have been taken by men who did not possess the Luther-like spirit of Garrison and his associates. But what if he had possessed a stronger spirit of love? or if he had been, as it were, a Luther-Melanchthon embodied in one soul? But God raised up a Luther and a Melanchthon, but no Luther-Melanchthon. . . . Before the war, how many wise and good men sought to prevent the sword from being drawn; and during the bloody struggle, how many cried out, in the words of the prophet [Jeremiah], 'O thou sword of the Lord, how long will it be ere thou be quiet? put up thyself into thy scabbard, rest, and be still,' to be answered only by the words of the same prophet, 'How can it be quiet, seeing the Lord hath given it a charge . . . ?' "

§4. THREE KIND BROTHERS-IN-LAW

By reason of Langstroth's inability, during long periods of ill-health, to make ends meet, he was at many times in need of pecuniary aid. Relatives stood by him and, so far as they were able, helped him out. Such assistance within the family would not, for the most part, become matter of record. There are, however, records of material help given him by three brothers-in-law, all of them men of character and substance. So far from expressing mere pity, their gifts seem to have come to him as tokens of respect.

The first instance is that of Almon Brainard of Greenfield (see Notes 5 and 6, Chapter VI), who gave him a home for six years (1852–58) while he was writing his book, revising the work for a second edition, and getting his improved hive on the market. It is possible that his young son also found a home with the

Brainards during some of that time, while the wife and daughters were remaining in Philadelphia.[3]

The means of permanent reunion for his family came to Langstroth in 1858 from his brother-in-law Aurelius B. Hull of Morristown, New Jersey, who gave him the home at Oxford, Ohio—a brick house and eight acres of land—where he lived continuously for the next twenty-nine years. He established an apiary there and resumed his observations and experiments. Oxford had been for a half-century a settled outpost of New England. It was a pleasant village for a man of intellectual taste, being the seat of Miami University and of Western College for women. The house that he dignified by his habitation belongs now to Western College and is used as a residence for teachers. It is called Langstroth House in his memory and has been little changed since his time.

From a third brother-in-law, William Gunn Malin (1801–1887) of Philadelphia, himself a collector of rare books, Langstroth received from time to time as gifts "nearly all" of his "old bee books." And Malin made bequests of $500 each to him, his two daughters, and a granddaughter. (See Note 7, Chapter VI.) Malin appears to have been a man of fine qualities. Langstroth wrote an account of his life which was published in *Gleanings* in October, 1895.

[3] He wrote in his Reminiscences: "I supplied the pulpit of the Congregational Church in Colrain, near Greenfield, the larger part of the time [from 1852] till the fall of 1857. My wife and daughters spent their school vacations in July and August with me in Colrain, where our son was living very near us, working on a farm. Oh those happy reunions!" *Gleanings in Bee Culture*, xxi, 206–7.

§ 5. EPISODE OF THE HEDDON HIVE

An article written by Langstroth and published by the *American Bee Journal* in the spring of 1888 brought upon his head a storm of expostulation. What he wrote was an out-and-out commendation of a type of hive devised by James Heddon, a beekeeper and manufacturer of Dowagiac, Michigan. His praise went so far as to imply if not even assert his conviction that Heddon's hive was a long step in advance of his own.

Heddon had invited him to come to Dowagiac, with all expenses paid, and see the hive in operation. He had been Heddon's guest for three weeks in April and had written and dispatched the article while there.

Its publication brought protests from men who were able to judge for themselves of the comparative merits of the two hives. Letters were printed lamenting the injustice which the writers thought that Langstroth had done to himself. His best friends were taken aback. Some of them, by now, had financial interests in the success of his hive, interests which they could hardly expect him to ignore. There were the Roots, for instance, manufacturers who had standardized the hive and its parts on the Langstroth principle and had made a considerable investment in machinery and materials.

And Charles Dadant & Son.[4] These patient men had now been waiting upon their friend Mr. Langstroth for seven years. In 1881 they had agreed to help him

4 See Note 7, Chapter VI.

with his long-delayed revision of his book—a demonstration of the excellence of his hive and his methods. He had seldom been able to work and had urged them to do more and more until gradually they had taken the task of revision upon themselves. Since he had no means of paying or indemnifying them, they had arranged with him and his former publisher, Lippincott, to publish the book at their own cost and risk and had agreed to pay him a royalty. Of late, yielding to appeals for money, they had been sending him checks in advance of publication. And now he comes out with a public statement which seems to imply that in his opinion the book is likely all too soon to become obsolete!

They wrote to him and remonstrated with him. He replied, assuring them that it gave him pain that they seemed to take his act so much to heart: "I value too highly your good opinion and friendship to do anything which ought to cause any coolness between us. . . . Let us, my good friends, leave this question to the decision of old father Time." He does not seem to have realized how he had hurt them until he got a chance to talk with the very man who had given him an introduction to them seven years before. Then he wrote them a letter, dated Cincinnati, May 10th, in which he said:

"I have had a long conversation with our mutual friend Mr. Chas. F. Muth, on this whole Heddon hive matter. It gives me great pain to find that he thinks that my endorsement of the Heddon hive comes out at a time that may in a business point of view affect adversely your pecuniary interests. *This is a view of the case that never occurred to me.*" (The italics are the author's. One can almost fancy Messrs. Dadant & Son in a whispered chorus, *That* is a view of the case that never occurred to *him.*) His letter went on: "Your loans of money came to me at a time when I was sorely in need of help, but I would much rather that my book should forever have gone out of print, than to have knowingly done anything to damage the interests of friends whom I respect and love."

The storm blew over.

Whatever the merits of Heddon's hive, it never even became a rival of Langstroth's and is now forgotten.

But Langstroth had no head for business.

XI

The Latter Years

THOSE two achievements which did the most to make Langstroth's life memorable, namely, his treatise on the honeybee and his perfection of the hive, were both realized by the time that he reached the middle of his forty-third year. He lived to twice that age. During these latter years, while beekeepers more and more were giving him credit for practical services to their industry, he did another memorable thing: he won their affection. The years were troubled with those ills and misfortunes which have taken up two chapters, but he came through them respected and loved.

When he took his reunited family beyond the Allegheny Mountains to Oxford in 1858 he set up an apiary on his place and resumed his study of the honeybee. An opportunity for a special service came to him in 1860, when the Italian bee was first brought to this side of the Atlantic. On that occasion he gave the patient and skillful help which has been described in Chapter VIII. Thereafter he took a special interest in the culture of this better race of bees. He promoted its distribution and domestication on this continent by studying how to adapt apiary practice to its distinctive ways and by the rearing, selection, and dissemination of queens.[1]

In the decade of the sixties, in this country and Canada, there was a great expansion, which might almost be called a boom, in the industry of beekeeping. To some extent, of course, it was due to the growth of population and the spread of agriculture westward. But there were contributing causes. With the movable-frame hive the care of bees had been made both easier and less liable to serious error. Langstroth's manual for beekeepers, the first soundly scientific work of its kind in this country, interesting to read and filled with all that a beginner needed to know, was in its third edition. And now one could stock an apiary with the new golden-banded bees, which had milder tempers than the blacks and brought home more honey. It was astonishing to learn how quickly a whole colony of blacks could be made over in complexion and character by giving it a fertile golden queen. All these contributions added to the fascination as well as the profit of the pursuit. And the one name more than all others that well-informed beekeepers came to associate with these better ways was that of the Reverend L. L. Langstroth.

With the growth of the industry came organization. Meetings of beekeepers were now reported by newspapers and agricultural journals. County and state fairs were bringing them together. State associations formed and began to overrun state lines. Out of them eventu-

[1] See the summary of his observations which concludes Chapter VIII; see also the first paragraph of A. I. Root's recollections near the beginning of Chapter XII.

ally came a national association, but not without fric-
tion. Among the promoters was here or there a man
interested not so much in the industry itself as in the
grist that it brought to his mill. The story is worth telling here because it shows Langstroth conducting himself creditably in a difficult situation.[2] His position was
unhappy because the movements toward national organization coincided with the Otis-King controversy
and became tangled with it.

In March, 1870, the Michigan association issued a
call inviting the beekeepers of America to meet in convention at Indianapolis on the 21st of December next.
The call was considered by two regional associations
which happened to meet simultaneously in September,
the Northeastern at Utica, New York, and the Northwestern at Decatur, Illinois. The Northeastern (H. A.
King, secretary) sent a telegram to the Northwestern
expressing a desire that the national convention be
held at Cincinnati "because this point is centrally located, is free from local influences, and is near the home
of Rev. Mr. Langstroth, whom we want present." But
before the message was received the Northwestern had
ratified the call for the Indianapolis convention and
had adjourned.

At Indianapolis in December beekeepers from eleven
states, the territory of Utah, and the Dominion of Canada filled the hall of the House of Representatives and

2 The account is pieced together from reports in the sixth volume of
the *American Bee Journal*.

organized the North American Beekeepers' Association.
R. C. Otis nominated the Reverend Mr. Langstroth
for president and he was elected unanimously. A letter
from H. A. King was read concerning a national convention soon to be held at Cincinnati. Robert Bickford
of Seneca Falls, New York, who had attended the
Northeastern meeting in September, was asked to tell
what had been done there about a rival convention at
Cincinnati.

He replied briefly, by saying that some of the members of
that Association manifested a *desire* to have the National Convention held at Cincinnati instead of at Indianapolis, but he
was not aware that any one was *authorized* to call such a convention. He did not know that such a call had been made until
he read it in Mr. King's paper. It was his belief that Mr. King
must have made the call *on his own responsibility.*

The North American association then appointed
three delegates to attend the Cincinnati convention and
effect a reconciliation if possible.

At Cincinnati in February, 1871, about a hundred
and fifty beekeepers from fourteen states and the Dominion of Canada met in Templars' Hall and organized
the American Beekeepers' Association. H. A. King nominated the Reverend Mr. Langstroth for president and
he was elected unanimously. Mr. Langstroth was present. He said that he would accept the office as a compliment, but only on condition that none of the active
duties devolve upon him, because his health would not

allow him to undertake them. The convention assented.

At the very beginning antagonism to this second association had excited a free debate in which Otis, as brusque as King was suave, made his presence known. The upshot of the debate was an agreement to merge with the North American association.

After the convention had settled down to the topic of beekeeping, King obtained a suspension of the regular order and proposed that a fund of $5000 be raised for Mr. Langstroth's benefit. He offered $50 to start with. The source, manner, and circumstances of this suggestion provoked indignant comment from several speakers, including William F. Clarke of Guelph, Ontario, and A. I. Root, but a committee was appointed to consider it and report. Langstroth held his tongue until a favorable report was read. Then he said that it was against his feelings to have his personal affairs taken up by the association; if his own wishes were to be consulted he would like the matter to go no further.

While the assembly was considering its proper business of beekeeping Langstroth was more than once asked respectfully for his opinion or advice. Some of his occasional remarks are extracted from the *American Bee Journal's* report and reprinted below, without change of the reporter's style. They give an idea of the range of his observation. They illustrate his way of speaking offhand. They afford a rare if not unique record of a part taken by him in a convention of beekeepers. The last paragraph of this extract indicates the respect that he won on this occasion, at least from the reporter:

Mr. Langstroth said that if there were no disposition on the part of the bees to swarm we should soon have an end of bees. He said no invariable rule could be laid down in regard to swarming. . . .

Mr. Langstroth said he had known foolish queens to put a multitude of eggs in the same cell. He had known queens to deposit eggs outside the cell, and that queen fertilized. . . .

Mr. Langstroth was requested to speak on the subject of the mel-extractor and its relation to bee culture. He said that in 1853 he became interested in the subject of extracting honey from the comb and using the comb for the bees again. He consulted mechanics. None of them helped him. If any one had said to him centrifugal force, he would have said eureka. A foreigner discovered the process. This discovery would again revolutionize bee culture in this country. Twice or thrice the amount of honey could be produced from the same stock of bees and the same care now as formerly without it.

Now, some means must be devised to disarm the public of the suspicion that the extracted honey was a manufactured concoction. The candying of honey was not an objection. Age did not hurt it. He tasted some twenty-five years old and it was good. He had good authority in saying that good honey was taken from the ruins of Pompeii, nearly two thousand years old. . . .

We have got to convince the public that this extracted honey was not adulterated. The way was to put the price down so that adulteration would be unprofitable. He thought the more the knowledge of how to manage this extraction and preserving of honey was diffused and acted upon, the better it would be for bee culture. . . . A vote of thanks was given Mr. Langstroth.

Mr. Langstroth added, in relation to young queens, that he had ascertained that the supposed enmity of bees to all unfer-

tilized queens was a mistake. He had put a very young unfer-
tilized queen on the opposite side of the comb on which a
fertilized queen was walking. A bee would sometimes stop and
stare at the intruder, as much as to say, "Does your mother know
you are out?" Sometimes they would hustle her out of the hive,
nearly killing her. Experiment only with *very* young queens.

Mr. Langstroth being called upon again said that he thought
the drone progeny of an Italian queen would be pure Italian
drones, be the drone by which the queen was fertilized a black
drone or an Italian drone. He said that when the Italian bees
were first introduced into this country there was opportunity
to test the theory. He said that in warm blooded animals where
there was a common circulation between the mother and the
unborn offspring, there was a decided influence exerted upon
the mother. Mares that had produced mules had years afterward produced horses with mulish characteristics and of mule-
like build. . . .

Mr. Langstroth said that there was every reason to believe
that the Italian bee was itself a hybrid. Long before the Egyptian bee was introduced into this country, there was evidence
of a bee in America with a tuft on the head like the Egyptian bee.
It was said, too, that the Italian bee could be produced from a
cross with a black bee.

In regard to the fertile worker, he said that Huber thought
workers had robbed a little and eaten of the ambrosia with
which the queen was fed. Then they might be bees produced
in imperfect queen cells, i. e., cells not quite large enough for a
queen, and a little larger than that in which a worker was produced. He said that instances had been found in which the head
of the bee was a drone and the anterior [*sic*] part a worker, and
vice versa. This was accounted for upon Mr. Wagner's theory of
the double germ. . . .

Mr. Langstroth said that it was possible for a worker grub
or egg to be cultivated by the bees and formed into a queen. . . .

Mr. Langstroth endorsed what Mr. Porter said about the
honey-producing plants generally. He said that there was no

honey at all in the buckwheat. He had gone over acres and acres of it, and had not seen a bee upon it. Again, he said the buckwheat was one of the best honey-producing plants. He had gone through acres of it and found it laden with bees. Much depended on climate, season and location; south of here it was worth little for honey. So with white clover. Some seasons it was good and some bad for honey making. The same was true of the goldenrod. He meant these remarks to show the different and contradictory observations that might be made from different standpoints, and to show the need of charity in comparing experiences. . . .

By a unanimous vote of the Association [Mr. Langstroth] was given the special privilege of speaking when he chose, and as long as he chose. He solved many knotty questions, and poured oil upon the waters when they were troubled.

The early seventies were unhappy years for him. His only son died in 1870, his friend Samuel Wagner in 1872, and his wife in 1873. The son's untimely death took away his strongest support. The young man's life had given much promise. A beekeeper of Seneca Falls, New York, Robert Bickford, wrote for the *American Bee Journal* an article in which he said:

Mr. James T. Langstroth was well known to most of the leading beekeepers of the country, either personally, or through business correspondence relating to bee culture, during the last ten years; and certainly no young man could have more completely won the confidence of all with whom he came in contact, than he has done, by his intelligence, modesty, strict integrity, promptness, candor, and perfect manliness in all his transactions. Aside from bee culture, he took an active interest in, and was generally at the head of, all patriotic, charitable, or social organizations in his immediate neighborhood. In fact, he was the leading young man in the town in which he lived.

But above all his other excellent qualities, stands, in my estimation, his unselfish and untiring devotion to his aged, infirm, and dependent parents. Next to the care of his own little family, his father's, mother's, and sister's comfort, wants, and wishes, were uppermost in his mind.

Langstroth's daughter and her husband, H. C. Cowan, came into the household during his wife's last illness and continued to live with him. In 1874 he sold his apiary and never afterwards had more than a few colonies at any time. His "head trouble" disabled him for long periods. In 1887 the household removed to Dayton, where Cowan's business required him to live, and there Langstroth spent his last eight years.

A few months before his death he recovered his health and spirits to an unusual degree. In September, 1895, he felt so well that he went to Toronto, accompanied by his daughter, Mrs. Cowan, and attended a convention of the North American beekeepers' association. He spoke there, by invitation, relating the story of his labors with an early shipment of Italian bees.

He died on Sunday, October 6, 1895, in the Wayne Avenue United Presbyterian Church at Dayton. The circumstances are told in the following extract from a letter which Mrs. Cowan wrote, a few days afterwards, in response to a request from the editor of *Gleanings*, who published it in the October number:

I can give you only a brief account of my father's last days. When asked, the Sabbath previous to his release, by our pastor whether he felt able to make the address at our communion service, he replied, "I shall be most happy to do so," adding, in response to the assurance that, if he did not feel able for it when the time came, he could be relieved, "Oh! I shall be able —it will be a joy to me, Mr. Raber. I am so glad you asked me!" He had been very bright and happy ever since his return from Toronto; but last week he took a heavy cold, and was much oppressed with it; and during the last few days he lost strength so rapidly, and seemed so feeble, that I wished him to notify our pastor not to depend upon his assistance on Sabbath. He was, however, confident that he could carry out his part in the services, and was so anxious to do so that I could not insist.

On Sabbath morning he was unusually bright, and overflowing with happiness and gratitude to the Lord for his blessings. My eldest son, with his wife and baby, had been spending a week with us, and he was much pleased with, and proud of, his little great-granddaughter. He asked her mother that morning to wheel her little carriage into his warm room, and I shall not soon forget how happy he looked as he sat beside it, talking to and caressing the little one. They were at the church.

After dressing, father seemed much fatigued, and I again asked him whether he thought it were best for him to try to preach. He replied, "Oh, yes! I will say a few words, and then I will come home and rest, rest, rest." . . .

Before [the] preaching, Rev. Amos O. Raber moved the pulpit to one side and placed a chair on the front of the platform. Father began to address the audience, sitting, with some explanatory remarks as to his weakness. After a few introductory sentences requesting the prayers of the congregation for himself and the service, he said: "I am a firm believer in prayer. It is of the love of God that I wish to speak to you this morning— what it has been, what it is, what it means to us, and what we ought——" As he finished the last word he hesitated; his form straightened out convulsively; his head fell backward, and in about three minutes he was "absent from the body, at home with the Lord."

His grave in Woodlawn cemetery at Dayton is
marked by a polished block of granite, an expiatory of-
fering in which many an American beekeeper had a
share. The stone is inscribed, sincerely if not with lap-
idary skill, to the memory of the

" 'FATHER OF AMERICAN BEEKEEPING'
by his affectionate beneficiaries in the art who, in remembrance of the services rendered by his persistent
and painstaking observations and experiments with the
honeybee, his improvements in the hive, and the charm-
ing literary ability shown in the first scientific and popular book on the subject of beekeeping in the United
States, gratefully erect this monument."

XII

Estimate and Eulogy

Soon after Langstroth's death, in its issue of December 15, 1895, *Gleanings in Bee Culture* published a symposium of articles in his memory, including contributions from beekeepers and others of the United States, Canada, Great Britain, Germany, and Switzerland. A selection of these articles is used to conclude this final chapter. Some of them speak from close acquaintance and friendship. On the whole they show the impression that Langstroth's character, talents, and personal qualities had made upon discerning and honest men who were his contemporaries.

The founder of *Gleanings,* Amos Ives Root, had known him well. Root had been a beekeeper since 1865 and since 1869 a manufacturer of beekeepers' supplies at Medina, Ohio. He is said to have done perhaps more than any other man in America to commercialize the beekeeping industry.[1] His standardized apparatus was introducing the Langstroth hive and system throughout large parts of the world. For the mid-October (1895) issue of *Gleanings* he wrote a page of personal

[1] By John I. Falconer in the *Dictionary of American Biography.*

recollections. It began with a reference to his first interest in bees, and continued:

During the whole of my busy life, perhaps no other hobby
has been pursued with the zeal and keen enjoyment that my
acquaintance with the honey-bees has. It seemed for a time as
if a new world were opening before me. After I had questioned
again and again everybody who kept bees, or knew any thing
about them in our neighborhood, I began impatiently ran-
sacking books and periodicals. The more I found, the more I
thirsted for deeper knowledge. I took a trip to Cleveland, prin-
cipally to overhaul the bookstores for works on bees; but I did
not dare to tell even the members of my own family that I was
taking such a trip by stagecoach (for it was away back in the
days of stagecoaches, before our railway was built), just to satisfy
my thirst and curiosity in this direction. I remember well how
the book-keeper pulled down his volumes one after another,
rapped the dust off, and began extolling their special merits. It
did not take me many minutes to decide that Langstroth's book
was *the* book. I was obliged to stay over night at the hotel, for
the stage made only one trip daily. I read and read, away into
the night; and it was during that night I commenced my acquaintance with the Rev. L. L. Langstroth. He told me just
what I wanted to know. My craze was not (certainly not at that
time) to make money, but rather to know more about God's
wonderful gifts—these strange and *curiously* wonderful gifts
which he has provided for the children of men. I did not look
at it then just as I do now; that is, I am sorry that, in those
earlier days, I did not recognize the Almighty as a loving father.
But Langstroth's book helped me a great deal, right in the line
where I sorely needed help. His wonderfully genial, friendly,
and sociable way of telling things enlisted my sympathies at
once.

I told you I was not studying then for the *money* there was in
it. Langstroth never wrote about bees or did any thing else because of the *money* there was in it. Through all his busy life,

he, at least at times, seemed strangely oblivious of the *financial* part. More of this anon.

After I arrived home it did not take me long to find out whether Langstroth was still living. I made the acquaintance, by letter, of Samuel Wagner; got hold of Vol. I. of the *American Bee Journal.* By the way, I wonder whether there is anybody living now who will enjoy reading the first edition of Langstroth and the first volume of the *American Bee Journal* as I enjoyed it then. Why, the very thought of those old days of enthusiasm makes the blood even now tingle to my fingers' ends.

As soon as I found out that Mr. Langstroth was living at Oxford, Butler Co[unty]., O[hio]., I commenced correspondence. Then I wanted the best queen-bee to start with that the world afforded. It was pretty well along in the fall, but I could not wait till spring, as some of my friends advised me to do. I soon learned to look up to friend Langstroth with such confidence and respect that I greedily read again and again every word I could find from his pen—even his advertisements and circular in regard to Italian bees. When the book was read through once I read it again. Then I read certain chapters over and over; and when summer time came again, and I had little miniature hives or nuclei under almost every fruit tree in our spacious dooryard, each little hive containing a daughter of that twenty-dollar queen, *then* I read Langstroth's book with still *more* avidity and eagerness, finding new truths and suggestions in it each time.

I think I met him first and heard him talk at a convention in Cincinnati.[2] He was a wonderful talker as well as writer,—one of the most genial, good-natured, benevolent men the world has ever produced. He was a poet, a sage, a philosopher, and a humanitarian, all in one, and, best of all, a most devoted and humble follower of the Lord Jesus Christ. His fund of anecdotes and pleasant memories and incidents was beyond that of any other man I ever met; and his rare education and scholarly accomplishments but added to it all. No one else I ever saw could tell

[2] The *American Bee Journal* (vi, 194) records that Langstroth and Root were both at the Cincinnati convention of 1871.

a story as he would tell it. A vein of humor and good-natured
pleasantry seemed to run through it all. I think he enjoyed
telling stories,—especially stories with good morals; and they
all *had* to have a good moral or they could not come from L. L.
Langstroth. Not only the play of his benevolent face and the
twinkle of his eye, but the motion of his hands as he gave emphasis to the different points in his narration, showed how thoroughly he entered into his topic. . . .

His last public talk to beekeepers, if I am correct, was the one
given at Toronto; and I felt anxious at the time that some short-
hand reporter might be at hand who could give all his words,
and even his little stories, just as he gave them to us then. Perhaps others did not enjoy this talk as I did, because they did not
know him as I did. Why, that history of long ago, telling of the
troubles, blunders, and mistakes in introducing the Italian bees
from Italy to America, should be handed down to coming generations. It should be embodied in some of the standard works
on bees, in order to secure its preservation. . . .

When quite a child I was greatly interested in reading the
life of Benjamin Franklin. When I first became acquainted
with Langstroth I could not resist the suggestion that he was
much like Franklin. The maxims of Poor Richard suggest the
thought. Mr. Langstroth was remarkably well read in ancient
literature. He was familiar with the writings of great men in
all the ages. It rejoices my heart now to know that he has been
remembered for many years at our national conventions, and to
know that he was even present with his daughter at the one that
occurred so short a time before his death. He never seemed to
have a faculty for accumulating property; but what is *millions*
of money compared with the grateful remembrance with which
Langstroth's name will be spoken in every civilized land on the
face of the earth?

Charles Dadant (1817–1902) wrote for *Gleanings* on
this occasion "a review of apiculture in old times and
in old countries" as a means of gauging Langstroth's

services. Although in some parts this article goes over ground already covered in this book, it is not out of place here. And it gives a glimpse of the life of Charles Dadant himself, a life of usefulness in the United States and in France and honored in both countries. He wrote as follows:

The men who, instead of destroying their bees, conceived the idea of domesticating them in order to get the crop of honey without so much work, made use of hollow tree trunks; after that they made hives of baked clay, of wickerwork, of straw, of cut stone, and of boards, etc. Unfortunately it was difficult to get the honey without destroying the bees. Besides that, the wax being greatly sought after, especially for use in churches, where no other kind of illuminating material was employed, its value prompted bee-men to suffocate their bees in order to rob them of their stores. I should add here, that, in the greater part of the countries dominated by Catholicism, this massacre of colonies was inevitable, for their laws forced the inhabitants of villages to furnish the churches so many hundredweights of wax every year. It was thus that the destruction of the richest apiaries became a lucrative business, as it is today in France, in Gatinais, near Paris, where there are professionals who make a business of laying out apiaries, and who are supplied regularly by persons who raise bees, not for their honey, but for the purpose of selling them when the hives are full.

In proportion to the spread of bee culture, their habitations were improved. Particularly was this the case in Greece, where a knowledge of their habits and the methods of culture was developed.

Della Rocca, in 1790, relates that the apiculturists in the Cycladean Archipelago, in Greece, used long hives of baked clay, which they placed horizontally through the thickness of walls which were made expressly for that purpose. The bottom of each hive was removable, and one could get the honey almost without disturbing the bees, and without being subject to the stings of the bees as they issued from the front end of the cylindrical hive. He showed also a board hive, the frames of which were upheld by means of small top bars or slats, under which, and attached to them, the bees built their comb.

Hives with movable frames were still in use in Greece in Della Rocca's time, made after another fashion; for Liger, in his "Rustic House," printed in 1742, shows the design of a hive, with entrance, made of wickerwork, and furnished with top bars from which were suspended the frames. The progress of Greece in apiculture need not surprise us when we remember that, 300 years before the Christian era, Aristotle had already published some descriptions of the habits of bees.

If we refer to the ancient writings on apiculture we find that movable frames were not used in other countries until much later. Those who suffocated their bees, or those who sold bees to those who did so, did not have, and do not now have, hives with more than one compartment. Others, finding this practice cruel, or desiring to preserve their colonies, placed surplus cases on top of the hive bodies; afterward, hives with several divisions, or "stories," "horizontally divisible." Afterward some were constructed with two vertical divisions; then with three. Huber, toward the end of the last century, in order to study the habits of bees, made what is known as the "leaf" hive, which one might open as he would the leaves of a book as it stands on end.

Finally in Germany, Dzierzon published, in 1846, a description of his hive with movable frames, which opened from behind, and the frames of which were supported simply by little bars.

In the same year 1846, Debeauvoys, in France, published a book in which he described his hive with movable frames which were removed at the side. I have already related in the journals how I became acquainted with Debeauvoys and his hive. It was in 1849 that I visited the Paris exposition, when I saw, at the end of the hall through which I was rambling, a magnificent comb of honey on top of a board hive. Without paying any attention to the rebuffs which I met in my efforts to get near the hive, I pushed through the crowd. The exhibitor, Mr. Debeauvoys, was not there. One of his neighbors was there with an artificial brooder, in which chickens were hatched every day. These little chicks were perched on top of the brooder, and looked quite forlorn in their seeming distress at finding themselves in the midst of such a scene. Their owner said to me that the beekeeper would be back soon. Sure enough, Debeauvoys, a rather heavy man, with pleasing figure and lively deportment, arrived in a few moments and explained to me his hive and its manner of manipulation. It was a frame hive, opening at the sides, and it lacked nothing to make it practicable except a space between the ends of the frames and the hive; for the frames, having no top bar longer than their width, were supported by lengthening the side pieces, thus forming feet, touching at the ends the boards forming the front and rear of the hive. To get them out it was necessary to separate them from the walls of the hive by spreading the latter a little by means of a thin chisel-shaped lever. This plan succeeded very well so long as the frames were not propolized to the inside of the hive; but after that, their removal was difficult, if not impossible, without breaking the frames; so, although the Debeauvoys hive may have met with some favor in France for several years, and although his book has passed through six editions, the hive was soon abandoned as impracticable. This hapless invention had one unfortunate effect; namely, that of setting French beekeepers against the use of movable-frame hives entirely, thus retarding progress.

Debeauvoys sold to me the second edition of his book for 45 sous. On returning home I tried to make some of his hives, and transfer some colonies into them. Following the advice of Mr. Debeauvoys I did the transferring by night. It is to be remarked that I received no injury from stings. In the mean time I succeeded nicely. For one or two years I was proud of my hives, and showed them to all who desired to examine their interior.

Unfortunately, two or three years after that, we had a very mild winter, not to say a warm one. The fields of rye headed out in January. My hives were filled with brood in March; then a succession of very cold days in April caused the destruction of the twigs of the trees, which were as fully grown as they usually are in June. Then there were seen, instead of green leaves, young shoots blackened by the frost, hanging from the branches which gave them birth. Being too busy at that time to give my bees the attention they actually needed in order to regain the lost ground, I lost them all; and from that time on I had nothing more to do with bees until my arrival in the United States, in 1863.

Having procured the works of Langstroth, Quinby, and King, I immediately perceived the immense superiority of the movable bottom, and the space between the ends of the frames and the inside of the Langstroth hive. There arose, at this time, a discussion concerning the Langstroth patent, as to the priority of the invention of the frame hive with a space between the frame and the sides. King pretended that Berlepsch, a German beekeeper, had anticipated Langstroth. But it was proved that Mr. L. had applied for a patent about six months before Berlepsch had invented his hive. Besides, the Berlepsch hive, although having been generally adopted in Germany and Italy, could not for a moment stand any comparison with that of Langstroth. Its bottom board is fixed; it opens behind, so that, if one wishes to see the front frame, he finds it necessary to take them all out. The frames of the Berlepsch hive, being taller than they are broad, its surplus frames are smaller; and, besides, they are very limited in number, as the bottom of the hive is not movable.

In spite of the ill will which certain owners of the German hives showed, a comparison between the two hives, the Langstroth and Berlepsch, being entirely to the advantage of the former, it has gained a footing in Europe. I am proud to have been the first to describe and recommend it in France, where its distinguishing features—a movable bottom, and frames with spaces between them and the sides of the hive—have been adopted by all advanced bee-men.

The name of Langstroth is known and revered, not alone in North America, but in France, Switzerland, Belgium, Italy, and even in Russia, where the French edition of "Langstroth Revised" has been translated into Russian, in which language it has reached its second edition. Many apiculturists having described the qualities of our lamented friend, it suffices me to say that my son and I are happy to have been deemed capable by him to put his book, which was so far in advance of the times at the date of its first publication, abreast with all that has been achieved since; and above all to have succeeded in spreading its renown in all countries where the English language is known, and where he is considered, as well as in the United States, as a superior man, distinguished for his intelligence, his knowledge, his disinterested and unceasing work directed toward apicultural progress, to which he had devoted his life.

A tribute came from Albert John Cook, who had formerly been a professor of entomology in Michigan State College and had published a successful guide for beekeepers. He was at this time professor of biology in Pomona College, California. He was later, in 1911, appointed state commissioner of horticulture by Governor Hiram Johnson and held that office until his death in 1916 at the age of seventy-four. His letter, written at Claremont, contains an intimate portrait of Langstroth as a many-sided man:

It was with keen regret and sadness that I heard, on October 10th, that our old friend and benefactor—Mr. Langstroth was the benefactor of every beekeeper—had passed to his reward the Sunday before. I had recently had two letters from him, which spoke of health, vigor, and strength. We all know of his long journey to attend the Toronto meeting; and a very dear mutual friend wrote me but a few days before that he had entertained Mr. Langstroth at his house, had taken him for a long ride, and that he seemed very bright and vigorous, and talked of his friends with so buoyant a spirit, and of affairs with such keen mental zest, that, though nearly at the eighty-fifth milestone, yet he seemed likely to see many days of usefulness and activity before he passed to the great beyond. Thus it was that the news came to me as a shock; and with the thousands of others all over our great country, I bowed my head and heart with grief and sorrow that I should see the kindly face and hear the sympathetic voice no more.

It is certainly true that the world has never had a beekeeper who was more widely and justly known, loved, and appreciated than our dear Father Langstroth. And so the world never lost an apiarist who will be more widely and sincerely mourned.

I had visited our beloved and venerable friend repeatedly at his own home, and had entertained him at my home on several different occasions. I grew to love him as a dear personal friend, to admire him as a man of very rare native ability and acquirements, to venerate him as a man of the loftiest Christian character. Few persons ever suffered more cruelly at the hands of unscrupulous, selfish, designing men, and yet his great loving heart seemed to harbor no thought of revenge or unkindness. He exemplified, in a manner rarely witnessed even among good men, Christ's words, "Love your enemies, and pray for them that despitefully use you and persecute you." Indeed, he was a rare example of one whose life was wholly permeated and glorified with the spirit of the Master.

There are many men who excel in some one line of work. They may have a genius at invention; they may be wonderful at painting word-pictures, and thus gifted in the art of exposition; they may possess great intellects, and thus become great scholars; they may have that beautiful equipoise of character that insures just judgments, and makes them broadly and grandly influential; but how rare to find these qualities united in one and the same person! Mr. Langstroth was all this. His keenness of vision as an inventor was remarkable; his power as an investigator and writer was graphically illustrated in his admirable work on the honeybee; his ability and scholarship were known, recognized, and appreciated by all who knew him; while his beautiful character, that thought no evil, could hardly understand or believe that others were selfish, calculating, and willing to take advantage of his unsuspecting nature. His was a great mind, his a true loving heart, he that noblest work of God, a true, sweet, Christian character.

Today we know positively that Mr. Langstroth was the inventor of the first practical movable-frame beehive. The German top-bar hive, with combs fastened to the side, was a previous invention, as was the close-fitting hive of Major Munn; but neither of these was known to him previous to his own invention, and each was as inferior to his as is the sickle to the self-binder. Mr. Langstroth had the vision to see a great need, and the genius to supply it; and in so doing he shared the honor and glory of very few men—that of revolutionizing a great industry, and changing entirely its methods. He did more than this: he did his work so well, that, though nearly fifty years have rolled by, yet no one has been able in all that time to improve upon his invention in any essential particular. What a compliment to him, that his hive, essentially as it was given to the world nearly half a century ago, is today the hive of nearly all our brightest and most successful beekeepers! No one can gainsay the fact, no one can deny the glory, of such an accomplishment. I can not find a parallel case in all the history of inventions. The sewing machine, the reaper, the steamboat, the railroad locomotive of today, could hardly claim relationship with the first ones given to the public; indeed, we are told that no one can afford to run a steamship of a score of years ago.

Mr. Langstroth was also an author of the highest type. His

"Honeybee" will ever remain a classic in bee literature. The incisive style, the pure vigorous English, as well as the fascinating subject matter, alike charm the interest and awaken the deepest admiration. Like another classic, Darwin's "Voyage Around the World," he opens up to us the secrets of investigation, and we are charmed as we discover how his mind worked its way upward in the realms of invention and scientific discovery, and it is equally true that his honesty was as thorough as was his genius at invention or his ability to describe. He was no plagiarist, either as a writer or inventor. Even the thought of claiming the work or thought of others was revolting to him. Had the same been true of others, Mr. Langstroth would have died a rich man. In all his writings he was overscrupulous and particular to give every possible credit to others. . . .

Socially Mr. Langstroth was very exceptional. Time always took flight when he became a companion. An hour or even more was all too short for the mealtime when he sat at the board; and the hours for sleep were crowded at both ends of the night when he was an inmate of the home. His wide reading, his knowledge of history, his acquaintance with men, his thorough knowledge of the Bible, and his practical adoption of its teaching and spirit, all combined to make him a delightful and most valuable and entertaining companion.

His lifework especially endeared him to beekeepers. His invention and discoveries were a special gift to every progressive beekeeper the world over. He personally contributed to the success and happiness of this entire class. No wonder that they loved him with genuine sincerity. No wonder that they mourn his loss with sincere sorrow. It is good that he could be present at the last meeting of the North American Association at Toronto. It will remain with us a pleasant memory that he died while administering the communion, and commending the love of the blessed Master, whose love came with such a rich effulgence in his own life, and through him spoke as a blessed benediction to all who knew him or came under his beneficent influence.

The Reverend W. F. McCauley, who had been Langstroth's pastor in Dayton but had lately taken another
charge in Toledo, wrote these recollections:

I was intimately acquainted with Mr. Langstroth for the last
eight years of his life, he being a member of my congregation
in Dayton for over seven of those years. It will be impossible,
in the limits allowed in this symposium, to do more than give
some general statements concerning this remarkable man.

He came to Dayton in August, 1887. I last saw him in September, 1895, on the occasion of his visit to Toledo, made while *en
route* to the Toronto convention. Twice in the period indicated
he suffered from his peculiar physical ailment and its resulting
melancholia. The first of these lasted about three years; the
second terminated not many months before his death. He derived some temporary help from a physician in the early part
of his residence in Dayton, and said joyfully, "I think I am fighting a winning battle." It did not so prove in the end; but we
all rejoiced that, after all, the close of his life was free from
shadow and gloom. The trouble referred to did not seem to
affect his intellect, but was of the nature of depression of spirit,
by which he was unfitted for his accustomed tasks. When in the
enjoyment of entire health no one could be more active or
companionable. . . . Geniality and earnestness were companion traits or ingredients of his nature, and it was delightful to
have an intimate association with him. Though he grew to be
nearly five years past fourscore, yet there was still much of the
dew of youth upon him: he was like a tree bearing simultaneously blossoms and ripened fruit.

He had a breadth of culture and intellect that marked him
as a man among men. He deserves the respect of all, not only for
his achievements along the line of bee culture, but for his general ability and high character. This is the estimate of a friend
who knew him in his various moods, and never found him other
than true and lovable.

He was frequently reminiscent, and would quote fine passages; or relate anecdotes, of such a character that the point illustrated would remain a long time in the mind. He was never unreasonable, but by nature was intense, and that quality endeared him to the writer. He uttered his convictions in unmistakable phrase, and gave others the same privilege. This enabled one to "get on" with him in famous fashion, for no time was wasted in false motions. Kindliness and straightforwardness, discretion and courage, and these dominated by heartfelt devotion to God and his truth, threw about him an atmosphere of light and warmth.

In all the foregoing there is no exaggeration; the effort has been to seek phrases that would convey a just conception of a character honored by one generation, and whose virtues should be preserved for the contemplation of another.

In the Y. M. C. A., in the pulpit occasionally, and in private life constantly, he sought to employ his strength to accomplish good. He was unselfish, and planned busily for the good of others. His life was broken into segments by physical affliction, but he did his best with his opportunities; and through the scattered clouds we can trace the shining arch to its base at the triumphant close of his days. . . .

A Canadian, William F. Clarke of Guelph, Ontario, shared in this tribute of respect. He had helped to found the North American beekeepers' association at Indianapolis in 1870 and had been at pains to attend H. A. King's rival convention a few weeks later at Cincinnati, where he took Langstroth's part and sought to discredit King. After Samuel Wagner's death in 1872 he had served for a time as editor of the *American Bee Journal*. He wrote as follows:

My first knowledge of Mr. Langstroth by name was in the early winter of 1863. I had engaged to edit an agricultural journal which was to be started in Toronto, January 1, 1864, to be called the *Canada Farmer;* and on surveying the field of my prospective duties it occurred to me that there was one branch of agriculture about which I knew nothing whatever—namely, beekeeping. I at once resolved to read up on the subject; and on making inquiries for the best works to peruse I met with "Langstroth on the Honey-Bee." I lost no time in plunging into its pages, which I found replete with interest. The book read like a fairy tale. I felt as if I had been introduced into a new world. Up to this time my knowledge of the bee did not stand beyond Dr. Watts' juvenile and moral song, which commences, "How doth the little busy bee"! I now felt that I must explore for myself the new world which had been opened up to me. Toward spring I corresponded with Mr. J. H. Thomas, of Brooklyn, Ontario, and bought of him a colony of Italian bees bred from stock he had obtained from Mr. Langstroth. This was my initiation into what Mr. W. Z. Hutchinson, as I think rightly, calls "the pleasant occupation of tending bees," the fascination of which is easier felt than told. If it were only for the gratification I have derived from this pursuit as a scientific pastime for more than thirty very busy years, I should feel that I owed Mr. Langstroth a large debt of gratitude.

I first met our lamented friend face to face at a beekeepers' convention held in Cincinnati during the month of February, 1871, of which he was elected president by acclamation. His health was poor then, and he accepted office as a compliment, on the condition that none of the active duties of the position were to be performed by him, as he did not feel that he had physical strength adequate to the task. The convention unanimously consented to this condition. I had a great deal of talk with him about the suit for infringement of his patent rights, in which he was then engaged. He had a deep sense of wrong, and felt most keenly the attempts which were being made to deprive him of what he believed to be his just dues. It cut him to the quick, that the very man who was doing the most to deprive him of the benefits of his movable-frame-hive invention should have placed him before the meeting as an object of charity by starting a $5000 subscription fund on his behalf. He said he did not want charity; all he asked was justice.

My intercourse with Mr. Langstroth at this time led me to form an exalted estimate of him as a man of high honor, scrupulous integrity, and unbending rectitude. The spirit showed by him toward those who were injuring him was admirable. There was no harshness, no display of unkind feeling, the predominant thought being to have his cause triumph because it was right. There was a lofty dignity and moral majesty about him which impressed me very deeply. I never met him again until in September last at the Toronto convention. He recalled the events that transpired at the Cincinnati convention, nearly a quarter of a century before, and was most profuse in his expressions of gratitude to myself and Mr. A. I. Root for the efforts made by us to have him righted on that occasion. It was a pathetic parting we had at the close of the [Toronto] convention, like that of the Ephesian elders with Paul. They sorrowed most of all that they should see his face no more; and I had the same feeling as I bade him adieu in Toronto, though I did not think the end would come so soon; nor did he. With improved health he was looking forward with almost youthful buoyancy of hope to doing some further work on which his heart was set. But it was not to be. He returned home to die, amid scenes the most hallowed and dear to him, and while engaged in the work he most sacredly loved.

"The weary wheels of life stood still at last." As a Canadian I am proud and glad that the last public tribute of honor and respect was paid to him, not only on our soil, but amid the classic surroundings of our educational department, where so many busts of departed greatness in literary and philanthropic walks of life are gathered. . . .

Though I met Mr. Langstroth on only the two occasions to which I have referred, I had correspondence with him at various times when circumstances arose which prompted it. When, at the request of the North American Beekeepers' Association, I took charge of the *American Bee Journal,* and removed it from Washington [3] to Chicago, his daughter, Mrs. Cowan, wrote me, at his request, a letter of approval and encouragement. . . .

Mr. Langstroth belonged to a class of beekeepers who are numerous in Great Britain, but comparatively scarce in this country, who engage in the pursuit, not so much for the money there is in it, as for the interest they feel in observing the nature and habits of these wonderful insects, and trying to uplift and ennoble the occupation as worthy a place of honor among intelligent and educated people. It is common in some quarters to despise and disparage this class of beekeepers; but for what reason I can not divine, so that it is almost necessary to defend the memory of this great and good man from the undeserved obloquy of not being a specialist in this line. He was one of those who do not believe money-making to be the all-important business of human beings in this world. Though he did not enrich himself by keeping bees, he performed services for others, the value of which is untellable. It is a poor return for these services which have done so much to ennoble beekeeping, both as a science and an art, to belittle the performers of them because they do not count their colonies by the hundred nor their gains by the dollar-and-cent standard. Practical beekeepers should hail those of the Langstroth class as allies and helpers, and be glad that any and all, according to their several ability, should aid the pursuit.

As a writer, Mr. Langstroth wielded a powerful and graceful pen. He was a master in controversy, and some of his articles of this character, that are on record in the earlier volumes of the *American Bee Journal,* are models of their kind. He did not fear to call a spade a spade. He dealt in no hollow compliments; and, while respectful and courteous toward all, he spoke the truth as he believed it, without fear or favor. Nothing low or vulgar ever marred his writings. There was a charm about his style that could not fail to interest all who read them.

---

[3] For several years before his death in 1872 Samuel Wagner lived in Washington, D. C., and published the *Journal* there.

Thomas G. Newman, the head of a publishing house in Chicago which had been (1875–1892) the publisher of the *American Bee Journal* and had brought out books dealing with bee culture, wrote a letter testifying, from his own observation, to the respect in which Langstroth's work was held in Europe. Newman's letter included what follows:

Father Langstroth's invention of the movable frame . . . was so perfect when announced that time and experiments have not improved it. While in different countries and climates its dimensions have been modified somewhat, yet the movable frame in use today, the world over, is substantially the same as when Father Langstroth introduced it to the world after he had privately experimented and severely tested it, in 1852.

That invention completely revolutionized the pursuit of beekeeping in all the civilized countries of the earth, and gave it such an impetus that it has now become one of the leading agricultural industries of the world. Its inventor's name will endure as long as bees are cultured, and will be remembered and revered by generations yet unborn.

His classic book, *The Hive and Honey-Bee,* stands at the head of beekeeping literature, and has been translated, in whole or in part, into the principal languages of the world; and on the natural history and habits of bees is a standard authority wherever beekeeping is an industry of importance.

Sixteen years ago, when I attended the bee and honey exhibitions in Europe, it was demonstrated to my observation that movable frames were almost universally used; and whenever the inventor's name was mentioned, it was cheered to the echo. In England, Scotland, Switzerland, Germany, Austria, and Italy, I found enthusiastic admirers of Father Langstroth, who truly loved him for his apicultural inventions and for his pure and benevolent character.

Unfortunately, Father Langstroth was not a millionaire; nor did he possess enough in his old age to secure even the necessaries of life. He had been too liberal and unselfish in his prime to even think of age or want. At a banquet in London, given in honor of the American representative to the bee conventions of Europe, a toast was proposed to the Rev. L. L. Langstroth for his apicultural investigations and his genius. Being called upon to respond to the toast, I referred to the misfortune of his financial state, and immediately, in true, large-hearted English style, the whole assembly rose to its feet and cheered his name, and a goodly contribution was then and there made and sent to Mr. Langstroth to cheer his heart and to help him during the approaching winter. There were present not only representative English and Scotch apiarists, but many from Continental Europe, and among these there were four or five editors of bee-periodicals. All were of one mind, doing honor to our loved American bee-master. . . .

At other bee conventions on the continent, similar scenes were enacted, particularly in Switzerland and Bohemia. At the latter, the Rev. Dr. Dzierzon forcibly stated his admiration of Mr. Langstroth, and sent a loving message to him by the writer. Among his other enthusiastic admirers I may mention the Baroness of Berlepsch, Augustus Schmidt, editor of the *Bienenzeitung*, Professor Sartori, Herr Vogel, Hilbert, and Carl Gatter.

Professor Butlerow, of Russia, Councilor of the Government, was the bearer from St. Petersburg to Prague of the imperial distinction of the Order of Santa Anna, by order of the late Czar, to confer the same upon the Rev. Dr. Dzierzon, for his apicultural research and inventions, and that dignitary was pleased to couple the name of Langstroth, of America, with that of Dzierzon, of Germany, and others, as the greatest men living in the apicultural world of the day. He then conferred the decoration on Dr. Dzierzon, with the usual ceremonies.

But space forbids further narration of the many interesting incidents within my knowledge, in proof of the topic assigned me. I will merely add that the educated and most prominent apiarists of the world, with one accord, attribute to Father Langstroth this well-earned honor: that in his life, character, and labors, he was one of Nature's noblemen—a modest, unassuming, honest man. He has fairly and fully earned the grand distinction of being the "Prince of Apiarists—the Huber of America"!

C. J. H. Gravenhorst, editor of the *Deutsche illustrierte Bienenzeitung,* wrote a letter saying that he had heard of Langstroth's death with sincere regret. He continued:

Every beekeeper in the Old and New World, who knows what this grand and noble American apiarian has done for advancing modern beekeeping, will feel as I do, and not the least my brother beekeepers in Germany.

Many years ago, when the late Samuel Wagner founded the *American Bee Journal,* I became a reader of and contributor to it, and in this way I learned what Langstroth had done and was doing to encourage beekeeping, especially when I received the first edition of his classic and renowned work, "The Hive and Honey-bee." I remember as if it were but yesterday when I received the book, and I would not lay it down till I had read the last word in it. How many times I have since read those wonderfully written chapters, the 11th, on the loss of the queen, and the 13th, on "robbing, and how prevented," and others. . . . Wherever one hears the best names of beekeepers spoken by the beekeeping world he will never miss the name of Rev. L. L. Langstroth.

Thomas William Cowan, editor of the *British Bee Journal,* wrote for this occasion a sketch of the history of efforts to improve the hive, noting the defects of certain early contrivances. He went on:

It was not till 1851 that Langstroth invented his frame hive, which, from its simplicity, cheapness, and practical adaptability to the purposes required, has conferred a lasting boon on beekeeping. There are no doubt some who think other methods are quite as good; but a very large and daily increasing number of beekeepers on this continent of Europe recognize that the principle introduced by Langstroth—and first published by him [in 1853] in his book on the honeybee—is the correct one. The opening of the hive at the top, the perfect interchangeability of the movable combs, and the lateral movement of the frames, have given the beekeeper the most perfect control over his bees, and have more than justified Langstroth's expectations when he wrote the note in his diary in 1851, that "The use of these frames will, I am persuaded, give a new impetus to the easy and profitable management of bees."

We here in Europe have for a long time held Langstroth in the highest esteem; have appreciated his invention, and only a few years ago we—British beekeepers—did ourselves the pleasure of electing him an honorary member of the British Beekeepers' Association, as a recognition for the services which he had rendered to apiculture.

A European leader in bee culture, Édouard Bertrand (1832–1918), citizen of the canton of Vaud in Switzerland, contributed to this symposium in a letter written at Nyon. Bertrand was the author of a work on the management of the apiary (*Conduite du rucher*) which had gone into numerous editions and become a classic. He had founded the *Revue internationale d'apiculture* and was serving as its editor at this time. His words are appropriate to the close of this book:

Through Mr. Dadant I had already received information of the death of Mr. Langstroth when your letter of Oct. 15 arrived. It was not without deep regret that I learned of the departure of that distinguished man to whom we owe so much; and I can assure you that, on this side of the Atlantic, the loss which the friends of the bee have just sustained will be no less keenly felt than in America; for Langstroth is considered everywhere, in Europe as well as with you, as one of the fathers of modern apiculture. Francis Huber, my fellow-countryman, prepared the way by discovering the secrets of the habits of the bees; and, fifty years later, Langstroth, in the United States, and Dzierzon and Berlepsch in Germany, crowned those efforts by giving to apiculturists systems of hives which have revolutionized the keeping of bees. But the manner in which the American inventor solved the problem of movable frames and the inspection of colonies, caused it to surpass the German method; and it is his hives and methods which have been adopted in the greatest number of countries, and which give the most brilliant results. I have, for my part, experimented with both systems; and without contesting certain merits in the Berlepsch model, I give the preference to the American hive, with loose bottom, and stores above.

But it is not alone for his useful invention that the memory of the great Langstroth deserves to be handed down to posterity. He has written an admirable book in which the elevation of the thoughts equals the extent of the writer's erudition as well as the richness of his observations, and which will remain the masterpiece of apicultural literature. Thanks to Mr. Dadant's translation, of which I am preparing a second edition, this work is now known to French-speaking apiculturists; and it has been produced in Russian through the labors of Mr. Krandratieff.

Our dear master's life had a glorious end, and one well worthy of it, as he died preaching the word of God.

Let us preserve his memory in our hearts.

APPENDIX

APPENDIX

Langstroth's Collection of Books about Bees and Beekeeping

REFERENCE has been made in the text to Langstroth's eagerness for books on bee culture. He acquired a respectable collection. After his death the *American Bee Journal* (in its issue of Nov. 28, 1895, vol. xxxv, p. 765) published a list of them. That list, though scant of detail and containing obvious errors, has served to identify almost every item and has been used as a guide in compiling the revised and expanded catalogue which is here appended.

The *Journal* avowed that its purpose in publishing the list was to help Langstroth's daughter, Mrs. Cowan, in finding a purchaser for the whole collection, and suggested that some state agricultural college would do well to buy it outright. But it appears rather to have been sold to a dealer and dispersed. A number of the volumes, lately discovered in the market, have been acquired for the Langstroth memorial collection in the Cornell library of beekeeping. This library now has eighteen volumes, bought in a single lot, which are shown by internal evidence to have been owned by Langstroth; it has twenty-six others, purchased at the same time, which correspond to editions that he is reported to have owned and may, some if not all of them, be the very copies that he possessed.

A list of what he owned has seemed to be worth publishing here for its own sake. Its publication will serve a secondary purpose if it brings other such appropriate books to this memorial collection.

[Symbols: *ABJ*, the *American Bee Journal's* list; L., Langstroth; asterisk (*), copy, now in the memorial collection, which L. is known to have owned; dagger (†), edition, now in the memorial collection, which L. is reported to have owned, and which may be the very copy that he had.]

† HILL [*or* HYLL], THOMAS: The profitable Arte of Gardeninge, now the thirde time sette forth: To which is added much necessrye matter, and a number of secrets with the Phisicke helpes belonginge to eche herbe, and that easie prepared. To this annexed, two proper treatises, the one entituled The marvelous governement, propertie, and benefyte of the Bees, with the rare secretes of the hony and waxe. And the other, the yearly conjectures, meet for husbandmen to knowe: englished by Thomas Hyll Londiner. [Part II has title:] A pleasant instruction of the parfit ordering of bees, with the marvelous nature, property & government of them; and the miraculous uses both of their hony and waxe (serving diversly as well inward as outwarde causes); gathered out of the best writers. . . . Whyche is now englished by Thomas Hyll, Londiner. [Title from ed of 1572.] London: Henry Bynneman. 1579.

HERESBACHIUS, M. C., *and* GOOGE, B.: Foure Bookes of Husbandrie, collected by M. Conradus Heresbachius, Councelor to the high and mightie Prince, the Duke of Cleve: containing the whole art and trade of Husbandrie. . . . Newly Englished and increased by Barnaby Googe, Esquire. At London, printed for John Wight. [Book IV has title: Entreating of Bees.] 1586.

BUTLER, CHAS.: The Feminine Monarchie: or the Historie of Bees. Shewing their admirable nature and properties, their generation, and colonies, their government, loyaltie, art, industrie, enemies, warres, magnanimitie, &c. Together with the right ordering of them from time to time: and the sweet profit arising thereof. London. Printed by John Haviland for Roger Jackson, and are to be sold at his shop in Fleetstreet, over against the Conduit. 1623. [*ABJ* indicates that L. had also copies of eds. of 1634 and † 1704.]

[Probably] MOUFET [*or* MUFFET], THOMAS: Insectorum sive minimorum animalium theatrum. . . . London. Printed by Thomas Cotes; sold by Benj. Allen. 1634. [Not certainly identified; *ABJ* has only: "Insectorum. By De Novert. 1634."]

HARTLIB, SAMUEL: Samuel Hartlib his Legacie: or an Enlargement of the Discourse of Husbandrie used in Brabant and Flaunders. . . . London. Printed by H. Hills for Richard Wodenothe at the Star under St. Peter's Church in Cornhill. 1651.

† RUSDEN, MOSES: A Further Discovery of Bees. Treating of the Nature, Government, Generation & Preservation of the Bee. With the Experiments and Improvements arising from the keeping them in transparent Boxes, instead of Straw-hives, as well to prevent their robbing in Straw-hives, as their killing in the Colonies. By Moses Rusden, an Apothecary; Bee-Master to the King's most excellent Majesty. Published by His Majesties especial Command, and approved by the Royal Society at Gresham Coll. London. Printed for the Author, and are to be sold at his house next the Sign of the King's Arms in the Bowling Alley, near the Abby in Westminster: And by Henry Million, Bookseller, at the Bible in the Old-Bailey. 1679.

WORLIDGE, JOHN: Vinetum britannicum; or, a treatise of cider, and other wines and drinks extracted from fruits growing in this kingdom. . . . Second impression, much enlarged. To which is added, a discourse teaching the best way of improving bees. [Half-title:] Aparium: or, A discourse of the government and ordering of bees. London: T. Dring. [Title from ed. of 1678; *ABJ* indicated that L. had a copy of the 3rd ed., 1691.]

† F[ERRIÈRE], DE LA, PRÊTRE [*and* MARQUIS]: Traité des Abeilles, où l'on voit la véritable manière de les gouverner & d'en tirer du profit. Avec une dissertation curieuse sur leur génération, & de nouvelles remarques sur toutes leurs propriétéz. Par M. D. L. F. Prêtre. À Paris, chez Claude Jombert, rue S. Jaques, au coin de la rue des Mathurins, à l'Image Nôtre-Dame. Avec Approbation & Privilège du Roy. 1720.

† GEDDE, JOHN: The English Apiary, or The Compleat Bee-master, unfolding the whole art and mystery of the management of bees, being a collection and improvement of what has been written . . . relating to this subject. . . . With a New Discovery of an excellent method for making Bee-houses and Colonies. . . . By John Gedde, Esq., approved by the Royal Society. London. 1721.

† WARDER, JOSEPH: The True Amazons: or the Monarchy of Bees. Being a New Discovery and Improvement of those Wonderful Creatures. Wherein is Experimentally Demonstrated, I. That they are all governed by a Queen. II. The Amazing Beauty and Dignity of her Person. III. Her extraordinary Authority and Power. IV. Their exceeding Loyalty and unparallel'd Love to their Queen. V. Their sex, Male and Female. VI. Their Manner of Breeding. VII. Their Wars. VIII. Their Enemies; with Directions plain and easy how to manage them, both in Straw-hives and Transparent Boxes; so that with laying out but Four to Five Pounds, in Three to Four Years, if the Summers are kind, you may get Thirty or Forty Pounds per Annum. Also how to make the English wine or mead, equal, if not superior, to the best of other Wines. By Joseph Warder of Croydon, Physician. The Sixth Edition. To which is Added, A Letter from the Author, concerning a late Treatise upon the Subject of Bees. London: Printed for John Pemberton, at the Buck and Sun, over against St. Dunstan's-Church, in Fleet-Street; and

John Osborn and T. Longman, at the Ship, in Pater-noster-Row. 1726. [*ABJ* listed also a copy of the 9th ed., 1765.]

SWAMMERDAMMII, JOANNIS AMSTELAED: Biblia naturae; sive historia insectorum, in classes certas redacta, nec non exemplis, et anatomico variorum animalculorum examine, aeneisque tabulis illustrata, insectis numerosis variorum naturae observationis. Omnia Lingua Batava, Auctori vernacula, conscripta. Accedit Praefatio, in qua vitam Auctoris descripsit Hermanus Boerhaave Med. Prof. Latinam versionem adscripsit Hieron. Dav. Gaubius. . . . Leydae. 2 vols: I, 1737; II, 1738.

THORLEY, JOHN, REV. (Oxon.): ΜΕΛΙΣΣΗΛΟΓΙΑ, or The Feminine Monarchy. Being an Enquiry into the Nature, Order, and Government of Bees, those Admirable, Instructive, and Useful Insects. With a New, Easy, and Effectual Method to preserve them, not only in Colonies, but in common Hives, from that cruel Death, to which their Ignorant, Injurious, and most Ingrateful Owners so commonly condemn them. A Secret unknown to past Ages, and now Published for the benefit of Mankind. Written upon Forty Years Observation and Experience. London: Printed for the Author, and sold by N. Thorley. 1744.

* MILLS, JOHN: An Essay on the management of Bees, wherein is shown . . . that the practice of saving their lives when their honey and wax are taken from them was known to the Antients, and is, in itself, simple and easily executed. London. 1766.

† WILDMAN, THOMAS: A Treatise on the Management of Bees; wherein is contained The Natural History of those Insects; With the various Methods of cultivating them, both Antient and Modern, and the improved Treatment of them. To which are added, The Natural History of Wasps and Hornets, and the Means of destroying them. Illustrated with Copperplates. London: Printed for the Author, and sold by T. Cadell, opposite Catherine-Street, in the Strand. 1768. [*ABJ* listed also a 2nd ed., 1770.]

* PINGERON [JEAN CLAUDE DE], Capitaine d'Artillerie & Ingénieur au Service de Pologne [1730–1795]: Les Abeilles, Poème

traduit d'Italien de Jean Rucellai, enrichi des notes historiques et critiques, et suivi d'un Traité de l'éducation de ces insectes. [Text and translation on opposite pages.] Amsterdam & Paris. 1770 [?].

* GÉLIEU, J. DE: Instructions pour les habitans de la campagne, contenant en abrégé la manière la plus simple et la plus sûre de gouverner les abeilles; extrait de l'ouvrage de feu J. de Gélieu, Mem. Soc. econ. Berne. (Part II.) [Mulhouse.] 1770.

SCHIRACH, ADAM: Wald-Bienenzucht, nach ihren grossen Vortheilen, leichten Anlegung und Abwartung . . . mit einer Vorrede, nebst des Herrn Verfassers Lebensbeschreibung begleitet von Johann G. Vogel. Breslau. 1774.

† KEYES, JOHN: The Antient Bee-Master's Farewell; or, Full and Plain Directions for the Management of Bees to the Greatest Advantage; disclosing Further Improvements of the Hive, Boxes, and Other Instruments, to Facilitate the Operations; especially that of SEPARATING Double and Treble Hives or Boxes, with Certainty and Safety, without Injuring the Bees; Interspersed with New but Important Observations: The WHOLE studiously adapted to GENERAL USE; with an appropriate method for the Curious. Also brief remarks on Schirach, and other distinguished Apiators on the Continent. Deduced from a series of experiments during thirty years. Illustrated with plates. By John Keyes, of Bee-Hall, near Pembroke. London: Printed for G. G. and J. Robinson, Paternoster-Row. † 1780. [*ABJ* listed also an ed. of † 1796; one such has same imprint as above; another of the latter year bears imprint: Dublin: Printed for P. Byrne, P. Wogan, J. Moore, and J. Rice.]

Scriptores rei rusticae. 4 vols. Ed. by Schneider. From Columella *De re rustica* [ca. A. D. 40]. 1787. [This date is that of *AJB;* the above title is of an ed. of 1794.]

ROCCA, L'ABBÉ DELLA: Traité complet sur les abeilles, avec une méthode nouvelle de les gouverner, telle qu'elle pratique à Syra, île de l'Archipel [Cyclades]; précédé d'un précis historique et économique de cette île. 3 vols. Paris. 1790.

* BONNER, JAMES: A New Plan for Speedily Increasing the Number of Bee-Hives in Scotland; and which may be extended, with equal success, to England, Ireland, America, or to any other part of the world capable of producing flowers. By James Bonner, Bee-Master, author of Practical Warping Made Easy, etc. Edinburgh: Printed by J. Moir, Paterson's Court: sold by W. Creech, Bell, and Bradfute, P. Hill, Mudie and Son, and by the Author, at Mr. Grant's, Leith Wynd, Edinburgh:—and by T. Kay, No. 332, Strand, London. 1795.

[BAZIN.] A Short History of Bees. In Two Parts. I. The Natural History of Bees, with Directions for the Management of Them, an Account of their Enemies, etc. from Réaumur, etc. II. An Ænigmatical Account of a Neighboring Nation—Their Queen, Her Palaces, Attendants, etc. London: Printed for Vernor and Hood. In the Poultry, by J. Cundee, Ivy-Lane, Newgate Street. 1800.

WILDMAN, DANIEL: A Complete Guide for the Management of Bees throughout the year; containing (1) A description of the Queen Bee. (2) The Generation of Bees. (3) Of the Drones. (4) Of the proper situation for a Bee-house. (5) The proper method of Swarming and Hiving. (6) Of separating the Honey and Wax. (7) Of the Enemies and Sicknesses incident to Bees. (8) Of feeding them in the Winter Season. (9) Explanation of the newly-invented Hives, with proper directions in what manner they are to be made use of. London. Printed for the Author, and sold by him at his house, Gray's-Inn Gate, Holborn. 1801.

* LOMBARD, M.: Manuel nécessaire au villageois pour soigner les abeilles. Avec figures. Paris. 1803. [1802, according to ABJ; the 1st ed. was of that year; the 4th, 1811, was entitled Manuel du propriétaire d'abeilles; there was a 6th, 1825.]

SPINOLA, MAXIMILIAN, MARCHESE: Insectorum Liguriae species novae aut rariores, quae in agro Ligustico nuper detexit, descripsit, et iconibus illustravit . . . Reimpr. Francofurti ad Moen. Jager. 1809.

VAREMBEY, J.: La Ruche française avec la manière de s'en

servir; ou nouveau procédé qui réunit les avantages de tous ceux publiés jusqu'à ce jour sur l'éducation des abeilles. Paris. 1811.

† HUBER, FRANÇOIS: Nouvelles Observations sur les Abeilles par François Huber. Seconde édition, revue, corrigée et considérablement ..ugmentée. [2 vols.] À Paris, chez J. J. Paschoud, Libraire, rue Mazarine, n° 22; et à Genève, chez le Même, Imprimeur-Libraire. 1814.

* GÉLIEU, JONAS DE: The Bee Preserver; or practical directions for the management and preservation of hives. Translated [by Miss S. Graham] from the French of Jonas de Gélieu, Late Minister of Lignières, at present Minister of the Churches of Colombier and Auvernier, in the principality of Neuchâtel; member of the Société Économique de Berne, etc. etc. etc. Published at Mülhausen. John Anderson Jun., Edinburgh, 55 North Bridge Street; and Simpkins and Marshall, London. 1829.

† THACHER, JAMES: A Practical Treatise on the Management of Bees, and the establishment of apiaries, with the best method of destroying and preventing the depredations of the Bee Moth. By James Thacher, M.D., Fellow of the American Academy of Arts and Sciences, etc. etc. Boston: Marsh and Capen, 362 Washington-Street. Press by Dow and Miles. 1829.

[HUBERT.] Fragments d'Hubert sur les Abeilles, avec une préface et une introduction par M. le Dr. Mayraux, Professeur d'Histoire Naturelle. Imprimerie de Béthune. Paris. 1829.

SMITH, JEROME V. C.: An Essay on the Practicability of Cultivating the Honey Bee, in Maritime Towns and Cities, as a Source of Domestic Economy and Profit. By Jerome V. C. Smith, M.D. Boston: Published by Perkins and Marvin. New York: J. Leavitt. 1831.

BEVAN, EDWARD: The Honey Bee; its Natural History, Physiology and Management. By Edward Bevan, M.D. London: Van Voorst, Paternoster Row. 1838. [The 1st ed. was pub. in London by Baldwin, Cradock and Joy, 1827.]

* NUTT, THOMAS: Humanity to Honey Bees; or, Practical

Directions for the Management of Honey Bees upon a new and improved plan, by which the lives of bees may be preserved, and abundance of honey of a superior quality may be obtained. By Thomas Nutt. Fifth Edition, revised, enlarged, and edited by the Rev. Thomas Clark. Wisbech: Printed by John Leach, for the Author, of whom it may be had at Moulton-Chapel, or at 131 High Holborn, London. Sold also by Longman and Co., Paternoster-Row, London; and J. Sholl, New York, America. 1839.

HUBER, FRANCIS: Observations on the Natural History of Bees. By Francis Huber. A new edition, with a memoir of the author, practical appendix, and analytical index. Illustrated with engravings. Cupar: Printed and published by G. S. Tullis. 1840. [Cf. the Nouvelles Observations, 2nd ed., 1814, above.]

WEEKS, JOHN M.: A manual or an easy method of managing bees, in the most profitable manner to their owner, with infallible rules to prevent their destruction by the moth. By John M. Weeks, West Farms, Salisbury, Vermont. New Edition. Revised and Enlarged. Boston: Weeks, Jordan and Co. 1840. [*ABJ* listed also an ed. as of 1857.]

* WIGHTON, JOHN: The history and management of Bees, with notice of a newly-constructed hive, by the author, John Wighton, Gardener to Lord Stafford. London: Longman and Co.; Norwich: Bacon, Kinnebrook, and Bacon, Mercury office. 1842.

* GÉLIEU, JONAS DE: Le Conservateur, ou la culture perfectionnée des Abeilles d'après les méthodes les plus récentes et avec application de celle de [Thomas] Nutt. Traité indiquant les moyens éprouvés par une expérience d'un grand nombre d'années pour conserver les ruches, pour les renouveller et en obtenir la plus grande récolte de miel possible. Avec 5 planches. Mulhouse, chez J. P. Risler, éditeur-libraire. 1843.

* MUNN, AUGUSTUS: A description of the Bar-and-Fram [sic] Hive, with an Abstract of Wildman's complete Guide for Management of Bees throughout the Year. London. 1844. [L. may have had a copy of an ed. of 1841 also.]

* HUISH, ROBERT: Bees, their natural history and general management: comprising a full and experimental examination of the various systems of native and foreign apiarians; with an analytical exposition of the errors of the theory of Huber; containing, also, the latest discoveries and improvements in every department of the apiary. By Robert Huish, F.Z.A. New edition, greatly enlarged. London: Henry G. Bonn. 1844.

GOLDING, ROBERT: The Shilling Bee Book, containing the leading facts in the natural history of bees, with directions for bee management. By Robert Golding, Hunton, Kent. London: Longman and Co., Paternoster Row. 1847. [*ABJ* indicates that L. had also a copy of the 2nd ed., 1848.]

* TOWNLEY, EDWARD: A practical treatise on humanity to honey bees; or practical directions for the management of honey bees, Upon an Improved and Humane Plan by which the lives of bees may be preserved, and abundance of honey of a superior quality obtained. By Edward Townley. New York: Printed by G. B. Maigne, 183 William Street. 1848. [This title has a marked resemblance to that of Thomas Nutt's book of 1839, above.—ED.]

* SCUDAMORE, EDWARD, M.D.: Artificial swarms. A treatise on the production of early swarms of bees. . . . 2nd. ed. London. 1848.

† MINER, T. B.: The American Bee Keeper's Manual; being a practical treatise on the history and domestic economy of the honey-bee, embracing a full illustration of the whole subject, with the most approved methods of managing this insect through every branch of its culture, the result of many years' experience. By T. B. Miner. Embellished by thirty-five beautiful engravings. New York: Published by C. M. Saxton, 121 Fulton Street. 1849.

TAYLOR, HENRY: The Bee-keeper's Manual; or practical hints on the management and complete preservation of the honey-bee and in particular in collateral hives. London. 1850. [4th ed. of a book first published in 1838.]

* MILTON, JOHN: The Practical Bee-keeper. . . . A new [2nd] edition, with frontispiece of bees and hives as exhibited by the author in the Crystal Palace. London. 1851.

† [FILLEUL, PHILIP V. M.] The English Bee-keeper, or suggestions for the practical management of amateur and cottage apiaries, on scientific principles. . . . By a country curate. London. 1851. [Also New York, 1851, The Cottage Bee-keeper, or suggestions, etc.]

RICHARDSON, H. D.: The hive and the honey-bee, with an account of the diseases of bees and their remedies. [A new ed., by J. G. Westwood. London. Ca. 1852.]

WOOD, J. G., REV.: Bees: their habits, management, and treatment. London. 1853. [*ABJ* indicates that L. had also a copy of an ed. of 1862.]

* Langstroth on the Hive and the Honey-bee. 1853. [Copy presented to L. by Dr. Joseph Beals, 1859.]

[?] FARRIÈRE, A. DE.: Les Abeilles. Avec 32 vignettes. Paris. 1855. [Not fully identified; *ABJ* has: "Des Abeilles. 1855."]

† *L'Apiculteur.* Bulletin mensuel de la Société central d'apiculture. Paris. T.-p. of Vol. I: | L'Apiculteur | praticien | journal | des | cultivateurs d'abeilles | marchands de miel et de cire | publié | sous la direction de M. H. Hamet | Professeur d'apiculture aux Luxembourg | 1856–1857 | Première Année. [Vol. II *omits* praticien. *ABJ* indicates that L. had Vols. I–VIII (1856–64) and XV (1870–71).]

ROUX, JEAN-FRANÇOIS: La fortune des campagnes. Traité pratique de l'éducation des abeilles. Lyon: Imprimerie d'Aimé Vingtrinier. 1856.

DEBEAUVOYS [CHARLES PAIX]: Guide de l'apiculteur. Cinquième édition. Revue, corrigée et augmentée de deux chapitres sur la Fécondation et sur les Combats des Reines, enrichie de nouvelles gravures. Paris [Libraire agricole de la Maison rustique]; Angers [Imprimerie-Libraire de E. Barasse]. 1856.

HAMET, H.: Petit traité d'apiculture, ou art de soigner les

abeilles (mouches a miel), contenant des notions succinctes de leur histoire naturelle, le gouvernement des essaims, l'emploi des ruches les plus avantageuses, la manière de faconner le miel, la cire, et l'hydromel. [Paris. 1856.]

LARDNER, DIONYSIUS: The Bee and White Ants, their manners and habits; with illustrations of animal instinct and intelligence. From "The Museum of Science and Art." With 135 illustrations. London: Walton & Maberly. 1856.

† PHELPS, E. W.: Phelps' beekeeper's chart; being a brief practical treatise on the instincts, habits and management of the honey-bee, in all its various branches. New York: A. O. Moore. 1858.

* HAMET, H.: Cours pratique d'Apiculture (Culture des Abeilles) professé au Jardin du Luxembourg. Ouvrage orné d'un grand nombre de figures dans le texte et du portrait de F. Huber. Paris: Aux Bureaux de l'Apiculture. 1859. [*ABJ* indicates that L. had also a copy of the 3rd ed., 1866.]

† HARBISON, W. C.: Practical Apiarian. Bees and Bee-keeping: a plain, practical work, resulting from years of experience and close observation in extensive apiaries, both in Pennsylvania and California. With directions how to make bee-keeping a desirable and lucrative business, and for shipping bees to California. New York: Saxton, Barker & Co. 1860.

BAUDE, JEAN: Traité d'apiculture pratique mis a la portée de tous les apiculteurs, et augmenté de nouvelles méthodes et observations. Lyon: chez l'Auteur. 1860.

SAMUELSON, JAMES, *and* HICKS, J. BRAXTON: The Honey-bee, its natural history, habits, anatomy, and microscopical beauties. London: J. Van Voorst. 1860.

† HARBISON, J. S.: The Bee-Keeper's Directory, or the theory and practice of bee culture, in all its departments, the result of eighteen years personal study of their habits and instincts. With an introductory essay by O. C. Wheeler, Corresponding Secretary of the California State Agricultural Society. Embellished

with eighty illustrations. San Francisco: H. H. Bancroft and Company. 1861.

† METCALF, MARTIN: A key to successful beekeeping: being a treatise on the most profitable method of managing bees, including the author's new system of artificial swarming, whereby all watching for swarms during the swarming period is done away with, and all loss by flight to the woods is prevented. New York: C. M. Saxton. 1862.

† [CUMMING, JOHN.] Bee-Keeping. By *The Times* Bee-Master. With illustrations. London: Sampson, Low, Son & Marston. 1864.

† HAMET, H.: De l'asphyxie momentanée des abeilles et des moyens de la pratiquer. Revised ed. Paris. 1864.

[?] Pratique complet d'apiculture, par un président d'un comité agricole du Finistère. 1864. [Not identified.]

† KING, N. H. *and* H. A.: The Bee-Keeper's Text Book, or facts in bee keeping, with alphabetical index, being a complete reference book, on all practical subjects connected with the culture of the honey bee, for both common and movable comb hives, giving minute directions for the management of bees in every month of the year and illustrating the nucleus method of swarming. [1st ed.] Cleveland, Ohio. 1864. Subsequent eds., 1867, † 1869 (9th rev. ed., 22nd thousand; New York: H. A. King & Co.), 1872, † 1878; also, King, A. J. (reviser): The New Bee-Keeper's Text Book, 24th ed., 52nd thousand, being a thorough revision of the Old Text Book by N. H. and H. A. King, enlarged and illustrated; New York: A. J. King & Co. 1878.

† KIDDER, K. P.: Kidder's Guide to Apiarian Science. . . . [Burlington, Vermont. 1858.] [*ABJ* indicates that L. had a copy of an ed. of 1865.]

* † NEIGHBOUR, ALFRED: The Apiary; or, bees, bee-hives, and bee culture: being a familiar account of the habits of bees, and the most improved methods of management, with full directions, adapted for the cottager, farmer, or scientific apiarian. London: Kent & Co.; Geo. Neighbour & Sons. * 1865. † 1878.

QUINBY, M[OSES]: Mysteries of Bee-keeping explained. Newly written throughout. Containing the result of thirty-five years experience, and directions for using the movable comb and box-hive, together with the most approved methods of propagating the Italian bee. New stereotyped and illustrated edition. [9th ed.] New York: Orange Judd & Company. 1865.

MONIN, F.: Physiologie de l'abeille, suivie de l'art de soigner et d'exploiter les abeilles. Avec gravures. Paris. 1866.

† SHUKARD, W. E.: British Bees: an introduction to the study of the natural history and economy of the bees indigenous to the British Isles. London: Lovell Reeve & Co. 1866.

PETTIGREW, A.: The handy book of bees. Edinburgh and London. 1870.

BEVAN, EDWARD: [Another ed. of] The Honey-bee, etc. Revised, enlarged and illustrated by W. A. Munn. London. 1870.

ADAIR, D. L. (Editor): Annals of Bee Culture. A bee-keeper's year book, with communications from apiarians and naturalists. Louisville, Kentucky. 1872.

The Bee-Keeper's Magazine. An illustrated monthly, devoted exclusively to bee culture. Editor: H. A. King. Vols. I–V (1873–77).

HUNTER, JOHN: A manual of bee-keeping. London: Hardwicke & Bogue. 1876 (2nd ed.). † 1879 (3rd ed.).

† ALLEN, JOHN [pseud. of the REV. OSCAR CLUTE]: The blessed bees. New York: G. P. Putnam's Sons. 1878.

COOK, A. J.: Manual of the Apiary. Chicago: Thomas G. Newman & Sons. 1879 (4th ed.). † 1881 (6th ed.).

QUINBY, M., *and* ROOT, L. C.: Quinby's New Bee-keeping. The mysteries of bee-keeping explained. Combining the results of fifty years' experience, with the latest discoveries and inventions, and presenting the most approved methods, forming a complete guide to successful bee-culture. With 100 illustrations, and a portrait of M. Quinby. New York: Orange Judd Company. 1879.

NEWMAN, THOMAS G.: Bees and Honey; or, the management of an apiary for profit and pleasure. Third edition. Chicago: Office of the *American Bee Journal*. 1882.

[DZIERZON, JAN]: Dzierzon's rational beekeeping; or, the theory and practice of Dr. Dzierzon, of Carlsmart. Translated from the latest German edition [of *Rationelle Bienenzucht*] by H. Dieck and S. Stutterd. Edited and revised by Charles Nash Abbott, Editor of the *British Bee Journal*. With numerous illustrations. London: Houlston & Sons. Southall: Abbott Brothers. 1882. [Not in *ABJ*, but the copy in the L. memorial collection contains marginalia believed to be in L.'s hand.]

The American Apiculturist. A journal devoted to scientific and practical beekeeping. Published monthly by S. M. Locke, editor and proprietor. Salem, Massachusetts. Vols. I (1883) and II (1884).

PHIN, JOHN: A dictionary of practical apiculture. Giving the correct meaning of nearly five hundred terms, according to the usage of the best writers. Intended as a guide to uniformity of expression amongst bee-keepers. With numerous illustrations, notes, and practical hints. New York: The Industrial Publication Company. 1884.

† BALLANTINE, WILLIAM, REV.: A practical treatise on bee culture. Illustrated. Sago, Ohio. 1884.

ALLEY, HENRY: The Beekeeper's Handy Book: or twenty-two years' experience in queen-rearing. Containing the only scientific and practical method of rearing queen bees, and the latest and best methods for the general management of the apiary. Third edition, revised and enlarged. Published by the author. Wenham, Massachusetts. 1885.

* CLARKE, WILLIAM F.: A bird's-eye view of beekeeping. Press of the *Canadian Bee Journal*. Beeton, Ontario. 1886.

COOK, A. J.: The bee-keeper's guide, or manual of the apiary (13th ed., 15th thousand). Lansing, Michigan. 1888.

* ROOT, A. I.: The A B C of Bee Culture: A cyclopaedia of

everything pertaining to the care of the honey-bee; bees, honey, hives, implements, honey-plants, etc. Facts gleaned from the experience of thousands of beekeepers all over our land and afterward verified by practical work in our own apiary. Medina, Ohio. 1884. [L. is believed to have had also a copy of an ed. of 1891.]

INDEX

INDEX

BY WOODFORD PATTERSON

The references are to pages. L. signifies L. L. Langstroth.

Library of Congress Cataloging in Publication Data

Naile, Florence.
   America's master of bee culture.

   First published in 1942 under title: The life of Langstroth.
   "Langstroth's collection of books about bees and beekeeping": p.
Includes index.
   1. Langstroth, Lorenzo Lorraine, 1810–1895. 2. Bee culture—United
States—History. I. Title.
SF523.L3N34 1976   638'.1'0924 [B]   76-12817
ISBN 0-8014-1053-3

www.ingramcontent.com/pod-product-compliance
Lightning Source LLC
Chambersburg PA
CBHW050040220326
41599CB00044B/7238